Practical Aircraft Electronic Systems

ALBERT HELFRICK

Embry-Riddle Aeronautical University
Daytona Beach, Florida

Prentice Hall Education, Career & Technology
Englewood Cliffs, New Jersey 07632

Library of Congress Cataloging-in-Publication Data

Helfrick, Albert D.
    Practical aircraft electronic systems / Albert Helfrick.
        p.   cm.
    Includes index.
    ISBN 0-13-118803-8
    1. Airplanes—Electronic equipment.   I. Title.
TL693.H45   1995
629.135—dc20   94-26978
                                                           CIP

Acquisitions editor: *Charles Stewart*
Editorial/production supervision: *Raeia Maes*
Cover design: *Doug De Luca*
Manufacturing buyer: *Ilene Sanford*

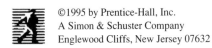

©1995 by Prentice-Hall, Inc.
A Simon & Schuster Company
Englewood Cliffs, New Jersey 07632

Printed in the United States of America

10   9   8   7   6   5   4   3   2   1

ISBN 0-13-118803-8

PRENTICE-HALL INTERNATIONAL (UK) LIMITED, *London*
PRENTICE-HALL OF AUSTRALIA PTY. LIMITED, *Sydney*
PRENTICE-HALL CANADA, INC., *Toronto*
PRENTICE-HALL HISPANOAMERICANA, S.A., *Mexico*
PRENTICE-HALL OF INDIA PRIVATE LIMITED, *New Delhi*
PRENTICE-HALL OF JAPAN, INC., *Tokyo*
SIMON & SCHUSTER ASIA PTE. LTD., *Singapore*
EDITORA PRENTICE-HALL DO BRASIL, LTDA., *Rio de Janeiro*

Contents

iii

Preface

This text is intended as a guide for aviation students who are involved in studies in aircraft-related fields. It could also serve as a self-study text for workers in the aviation industry to enhance their understanding of aviation electronics. The text handles all aspects of airborne electronic systems from an operational standpoint and discusses system operation to the block-diagram level.

Chapter 1 is an introduction to aviation electronics; it covers terminology and presents the applicable government and advisory groups. Historical notes are presented to gain an understanding of how some aviation electronics systems evolved to their present form.

Chapter 2 reviews electrical circuits and components. The chapter assumes some electrical circuit knowledge from a college-level or advanced high school physics course. The chapter covers radio-wave propagation and antennas, because many navigation systems rely on radio propagation.

Chapter 3 covers ac and dc electrical systems. Every aspect of the electrical system is discussed, including generators, batteries, wires, connectors, and bus systems.

Communications systems are covered in Chapter 4. This chapter introduces such general concepts as the superheterodyne receiver and frequency synthesizer and provides more specific information on radio-wave propagation. In addition to radio communications, microphones, satellite communications, and aircraft interphone systems are discussed.

Radio navigation systems are covered in Chapter 5. Historical information is provided to give the student insight into how some radio navigation systems evolved. All navigation systems in use in the United States are covered: nondirectional beacons, VOR, DME, LORAN-C, Omega, and satellite navigation. We include only the internationally recognized en route navigation systems and thus do not cover Decca, Consol, or Consolan, which are used only in Europe. Complete coverage of satellite navigation is provided due to its potential to become a major worldwide navigation system.

In Chapter 6, the entire instrument landing system is covered in detail: localizer, glide slope, marker beacons, compass locators, and radio altimeters. Microwave landing systems are also discussed, including precision DME.

Aircraft surveillance systems are the subject of Chapter 7. They include the air traffic control radar system and the secondary radar transponders. Both the older mode A/C transponder and the latest mode-S transponders are covered. In addition, a complete coverage of the collision avoidance system, TCAS, is provided.

Chapter 8 covers all types of displays, from the original mechanical instruments to the latest electronic displays. Because gyros are an integral part of the more sophisticated displays, this chapter also provides gyroscopic fundamentals.

Chapter 9 is an introduction to automatic flight control, often called autopilot. Only basic autopilot operation is covered, since this subject could comprise an entire book.

Chapter objectives preceding each chapter are provided to guide the student, and review questions for self-study or homework assignment are given at chapter ends.

Albert D. Helfrick

1

Early Development of Aviation Electronics

CHAPTER OBJECTIVES

This chapter covers the history of aviation pertinent to avionics development. The student will be introduced to the major regulatory and advisory bodies throughout the world and their activities. An understanding of the regulations that most affect aviation electronics systems will be gained. Aircraft nomenclature is discussed.

INTRODUCTION

Like the development of the airplane, the development of flight instruments and aircraft navigation equipment has been an evolutionary process. Keeping aircraft aloft was only the beginning of that process. Early pilots had to master the art of making turns and developing endurance. Records were set for altitude and long-distance flights and, most importantly, flying became safer.

In the early stages of aviation history, pilots and their aircraft required calm winds for flight. As aircraft technology developed, a visible horizon was sufficient for safe flight. But as the endurance of aircraft increased, the importance of distant weather conditions also increased. The horizon at the departure end of an air trip may have been visible, but the increasing speed and endurance of the airplane made the weather at the far end of the trip an important factor. The aircraft industry could not grow if the airplane was limited by the weather along a 3000-mile (mi) path. The ability to fly in all types of weather and, primarily, the capacity for zero-visibility flying were a necessity.

Zero-visibility flying requires three navigation systems or instruments. First, there must be a way of aligning the aircraft with the horizon. It was a well-known fact, even in the early days of aviation, that without a reference to the horizon the pilot of an aircraft could not keep an aircraft erect. By simply aligning the forces of gravity through the center of the body, it is easy to keep the body erect without reference to the horizon when standing on the ground. Obviously, most people can close their eyes and not topple over. But when accelerated motion is involved, as in making a turn in an aircraft, additional forces, such as centrifugal force, come into play. These additional forces cannot be separated from the forces of gravity when there is no way of knowing that the other forces are involved. When there is zero visibility, the pilot of an aircraft will attempt to align the forces acting on his or her body through the center of the aircraft, not knowing whether these forces are due only to gravity or to an accelerated turn. What may appear to the pilot to be an erect aircraft may in fact be an aircraft in an ever-tightening turn.

The second requirement for zero-visibility flight is a reliable method of measuring altitude. Without ground reference, there is no way for the pilot to know the aircraft's altitude without instruments or to be sure of a safe altitude for terrain clearance.

The third requirement for zero-visibility flight is a method of navigation that does not require ground contact from the departure point to the destination point. Most early long-distance flights were performed with maps, which would, of course, be of no value under zero-visibility conditions. The greatest precision of the navigation system is required at the terminus of the flight, during landing. To make a landing, the aircraft must be guided to the runway with the proper heading.

The first requirement for zero-visibility flight is met by the *artificial horizon,* an instrument with a blue and brown disk that is stabilized with a gyroscope. The disk appears as a miniature horizon, and the pilot aligns a small symbolic aircraft outline so that it is parallel to the artificial horizon. Another gyroscopic instrument used in the early days of instrument flying was the *turn-and-bank indicator.* This device indicated the rate of turn and could show when the turn was coordinated, that is, when the sum of all the forces acting on the aircraft was directed through the center of the aircraft.

The *sensitive altimeter* satisfied the second requirement for zero-visibility flight. To measure the altitude, early altimeters used the changing atmospheric pressure that occurs with increasing altitude. Although the principle was sound, early altimeters did not have the accuracy required to fly in near-zero-visibility conditions. If the visibility was less than the accuracy of the altimeter, the consequences were often catastrophic. The sensitive altimeter, which was developed in response to these problems, had an accuracy of 100 feet (ft) or so.

The third requirement was satisfied by *low-frequency radio transmission* and an *aircraft-mounted receiver.* Early instrument flights did not have a large network of navigation stations, and a system of radio-guided landing was considered to be more important. Navigation for short en route distances was accomplished with a method of projecting the flight path by knowing the ground speed, the angle of flight, and the time en route, a method known as *dead reckoning,* but the precision needed to make a landing required radio-assisted navigation. A radio beam was transmitted and the receiver in the aircraft indicated whether the aircraft was to the left, to the right, or directly on the beam.

It was with this equipment, the artificial horizon, a sensitive altimeter, and a radio landing guidance system, that Lieutenant James H. Doolittle made the first instrument take-off, flight, and landing on September 24, 1929. For the demonstration flight, a Consolidated NY-2 military training plane was selected because of its stable handling characteristics and its great strength. The aircraft was set in a glide path and literally flown into the ground, the only warning of the ground's approach being the touching down of the landing gear. The aircraft was outfitted with a sensitive altimeter made by hand by Paul Kollsman, whose company went on to make thousands of altimeters for commercial and military aircraft. The artificial horizon was made by the Sperry Gyroscope Company, which until its sale to the Honeywell Corporation in 1984 continued in the aircraft control field.

The radio equipment found in Doolittle's plane was made by Aircraft Radio Corporation, a division of Radio Frequency Laboratories. This company eventually became a part of Cessna Aircraft, Inc., and was later sold to the Sperry Corporation. Sperry eventually became another part of the Honeywell Corporation and no longer produces avionics under the Aircraft Radio name. Doolittle's demonstration flight was an important step for the future of three aircraft navigation and control industries and marked the beginning of the air transport business. Freedom from interruption due to weather conditions is an important part of air travel.

Although Doolittle's test flight proved that zero-visibility flying was feasible, many further improvements were required before the system would be suitable for airline and air transport use. One of the bigger problems with the system used by Doolittle was that the aircraft was set for a constant glide angle and literally flown into the ground. This was possible for the sturdy Consolidated bi-wing, but would fall short of being acceptable for an airliner filled with passengers. The sensitive altimeter provided the altitude relative to sea level and did not indicate the aircraft's height above the ground. If a chart of the local area was available, the sea-level altitude could be converted to height above ground, but without some sort of radio locating system, the position over the ground would not be known.

The intent of the Doolittle test flight was to demonstrate blind takeoff and landing. The total duration of the flight was about 15 mi, which is not typical of the distances required of an en route navigation system. For air transport applications, a viable long-distance en route navigation system is required. *Dead reckoning,* the technique of flying only with reference to a compass and with use of careful timing of checkpoints, was not accurate enough to guide aircraft close enough to the destination airport so that they could intercept the radio landing system installed there.

One of the first en route radio navigation systems installed in the United States was the four-course or A–N range. The four-course range operated in the low-frequency portion of the radio spectrum and used four or more transmitting towers having several hundred watts (W) of output power. Two antenna patterns were superimposed, one of which transmitted the Morse code letter A and the other the letter N. Where the two antenna patterns overlapped, the Morse A and N merged to become a continuous tone. Hearing the letter A or N was a signal to the pilot that he or she was off course and that an appropriate correction was necessary. No special equipment other than a receiver capable of covering the frequency of the range station was needed in the aircraft. The major disadvantage of the range was that only four courses could be defined with the system.

By mounting radio direction-finding equipment in the aircraft, omnidirectional radio facilities could be used by aircraft by homing to the beacon transmitters. Airways could exist between any two beacon transmitters that were within radio reception range. The development of airborne radio direction-finding receivers took place around 1940, and nondirectional beacons were constructed in large numbers.

With the exception of those periods during which an aircraft passes directly over a low-frequency navigation station and the pilot is in the *cone of silence,* the low-frequency navigation aids do not give any indication of position. Beginning about 1939, marker transmitters were installed for periodic fixes along the airways structured around low-frequency A–N ranges and nondirectional beacons. Indications of these en route markers, which were low-powered transmitters operating at 75 megahertz (MHz), were received by the aircraft when the aircraft passed directly overhead.

Before World War II, a complete en route navigation system was developed in the United States based on the low-frequency navigation aids and the 75-MHz marker beacons, but all low-frequency navigation systems suffer from several serious problems. One is *skywave interference,* often called "night effect." This occurs when signals arrive at the aircraft by ground-wave and ionospheric propagation. Another problem is *static* from thunderstorms. This problem was particularly serious because, when the weather was bad and the radio navigation system was most needed, it was most likely to be degraded by static.

After the war, the low-frequency navigation system was not expanded while a search was underway for a replacement that would offer more flexibility and would not be susceptible to the problems of the low-frequency radio navigation aids. In 1946, a very high frequency (VHF) omnirange system of navigation, or a VOR system, was developed. This system operated in the VHF spectrum, where there is no ionospheric propagation and where the effects of static are much less. In addition, the VOR system allowed aircraft to fly any radial course desired to or from the VOR station. The VOR was very successful and many VOR installations were made between 1946 and 1949. For all-weather landings, the localizer and glide slope systems were developed and installed at larger airports. TACAN, which is an improved omnidirectional navigation system, was installed in the mid-1950s and is intended primarily for use by military aircraft. The distance-measuring (DME) portion of the TACAN system found increasing use during the late 1950s and early 1960s to the point that it is now required for high-altitude aircraft.

Through the decades of the 1960s and 1970s, VOR, glide slope, localizer, and DME were the mainstays of civil aviation navigation in the United States. It was during this time that airborne equipment made great strides. Heavy vacuum-tube circuits made way for lighter transistor equipment and, later, even lighter equipment that used integrated circuits. Advances were made in improving the ground stations, while airborne equipment manufacturers improved the airborne systems. DME and weather radar, systems usually found in only larger aircraft, became common equipment in civil aircraft. The invention of the area navigation computer greatly improved the versatility of an already highly useful VOR system.

After nearly 40 years of avionics activity, the FAA decided in the 1980s that the U.S. airspace required modernization. VOR has remained as the standard en route navigation system, although avionics manufacturers are moving toward more sophisticated area navigation computers and airborne LORAN-C and satellite-based navigation systems. The microwave landing system is being installed to allow improved landing at airports already certified for instrument landing system (ILS) approaches and to allow instrument approaches at airports that have no ILS facilities. An improved air traffic control secondary radar system, the mode-S system, is being installed to enhance air traffic control and safety. A collision avoidance system based on the mode-S system was developed to reduce the chance of midair collisions in heavy air traffic.

THE USE OF AVIATION ELECTRONICS

To help understand how modern aviation electronics equipment is involved in the flight of aircraft, consider a hypothetical trip from a medium-sized airport to a small field where the aircraft will be refueled, take off again, and finally arrive at its destination, a large city airport.

The hypothetical trip begins in the morning at County Airport, which has a federally operated control tower. County Airport serves a small city and has corporate aircraft based at the field and a commuter airline service. Because of the commuter operations, County Airport has an instrument landing system (ILS), which allows flights to be conducted during poor visibility conditions. Before starting the engines, the pilot tunes the communications radio to the automatic terminal information service, or ATIS, to hear a recorded

Photo 1-1   A complete set of avionics for general aviation. Photograph courtesy of Bendix Corporation.

announcement concerning the local weather, the winds, and the active runway. After starting the engines, the pilot contacts the ground controller, using the same communications radio, or COM radio, and requests permission to taxi to the active runway. Permission is granted by the ground controller, and the aircraft is taxied to the active runway, during which time the COM radio remains on the ground control frequency so that any further information may be received from the ground controller. After the aircraft has reached the active runway and the preflight check has been completed, the pilot uses the COM radio to tune to the tower frequency, contacts the tower, and requests permission to take off.

Permission is granted and the aircraft taxies to the runway, applies full power, and takes off. During the takeoff roll, lift off, and climb out, and until the aircraft is approximately 5 miles from the airport, the COM radio is left on the control tower frequency.

After the aircraft has reached an altitude of about 1000 ft above the ground, the navigation receiver, or NAV radio, is tuned to a VHF navigation station, or VOR, and the aircraft is turned in the direction of the VOR station. The flight progresses to the navigation station. Our hypothetical aircraft is equipped with distance-measuring equipment, or DME, that gives the distance to the VOR station, which, for our imaginary trip, is 40 nautical miles (nmi).

Once clear of the airport traffic area, at about 5 nmi, the pilot can monitor an emergency frequency or any of the universal communications frequencies, called *unicoms,* or simply turn down the volume of the communications receiver and not listen to anything.

The cruising speed of the aircraft is 180 knots, which means that at 40 nmi the VHF navigation facility, or VOR, will be reached in about 13 minutes. After a few minutes, the DME has calculated a ground speed of 185 knots, which is a bit higher than the airspeed reading, indicating that there is a small tailwind.

After reaching the VOR, the aircraft takes a new heading and flies away from the VOR toward a small uncontrolled field where the pilot will land and take on fuel. *Uncontrolled* means that the airport does not have a control tower. This condition is typical of small fields, where the volume of traffic does not warrant a tower. The field is 38 nmi from the VOR, and this gives the pilot some time to use the COM radio to contact a flight service station, or FSS, to obtain weather information pertinent to the area around Small Town Airport. In this case, the pilot calls the FSS on the COM radio but listens for an answer on the VOR signal. This method is called *duplex* and is implemented to reduce interference.

Although the VOR and DME could be used easily to find Small Town Airport, the airport has a low-power nondirectional beacon, a low-cost navigation system suited to small airports, on the field, and the pilot decides to use the aircraft's automatic direction finder, or ADF, and home directly to the field from a few miles out. Although Small Town Airport does not have a control tower, it does have communications, and the pilot calls Small Town on one of the unicom frequencies and asks for the number of the active runway. The pilot keeps the COM radio tuned to Small Town's unicom, where other aircraft heading to Small Town or in the airport's traffic pattern are making known their positions and intentions. After a few minutes homing on the nondirectional beacon with the ADF, Small Town Airport is in sight, and the pilot enters the traffic pattern for the active runway and announces over the COM radio that the traffic pattern has been entered. Because Small Town does not have a control tower, the pilot will not receive an acknowledgment of the transmission from the tower, but may receive an acknowledgment from other aircraft in the vicinity of the airport that he or she is in sight. There is no tower in control and it is up to each pilot to assure a safe separation.

Once on the ground at Small Town, the same unicom frequency is used for directions to the fuel area and to summon the fuel truck to the aircraft for the fuel delivery. After receiving fuel, the aircraft is taxied to the active runway, the COM radio is used again to announce the departure, and the aircraft leaves Small Town Airport en route for the last leg of about 100 nmi to Big City Airport. Big City is a busy airport, has a control tower, and, in addition, is within class B airspace, in which all aircraft must contact the approach control and be equipped with a radar transponder.

Immediately after leaving Small Town Airport, the pilot tunes to a VOR facility near Big City Airport and enters the location of Big City into the area navigation computer, or RNAV. The RNAV computer gives the pilot steering information to Big City, and the DME will indicate the distance, ground speed, and required time to reach Big City. During the 100-nmi trip from Small Town to Big City, the pilot will check the en route weather by tuning a low-frequency beacon on the ADF and requesting additional information from an FSS. When the aircraft is about 5 nmi from the edge of the class B airspace, the pilot will set the COM radio to the frequency of Big City's ATIS and receive important information concerning the weather, winds, and active runway at Big City. Before entering the class B airspace that surrounds the airport, the pilot will contact Big City approach control, stating his or her position and destination—in this case, Big City. The controller at Big City approach control will instruct the pilot to dial a number into the radar transponder and to identify, which means to press the ID button on the transponder.

The approach controller will acknowledge to the pilot that radar contact has been established and will instruct the pilot on course changes. For the hypothetical flight to Big City Airport, the controller will allow the pilot to fly, still using the RNAV, directly to the airport under radar contact.

When the airport is in sight, the approach controller will instruct the pilot to contact the tower, which will clear the aircraft to land. Once on the ground, the pilot will contact the ground control for clearance to taxi to a tie-down area, and the airplane trip has concluded.

Our hypothetical trip was made under visual flight rules, VFR, which control the conduct of aircraft where visibility is good; typically, the pilot has ground contact at all times. When visibility is poor, instrument flight rules, IFR, are used. IFR requirements are for more equipment, both in the aircraft and at the airports, and more pilot training. There are a large number of procedural requirements relative to air traffic control and filing flight plans.

The intent of this hypothetical flight from County Airport to Big City Airport via Small Town Airport was to illustrate the types and uses of aircraft electronics equipment. The communications transceiver is the most heavily used piece of airborne equipment and, for that reason, aircraft are often outfitted with two communications transceivers. An in-flight failure of a communications transceiver, while not necessarily a dangerous situation, would be a hardship on most flights.

The VHF navigation set is the next most frequently used piece of equipment, and it, too, is often duplicated on aircraft. The ADF was used for navigation at the small airport and could be used as a back up navigation system for the VOR.

The radar transponder was mandatory for flight into the class B airspace surrounding Big City Airport, and, quite simply, an aircraft cannot land or take off from Big City Airport without a transponder. Even though the transponder does not provide any pilot readout to make the task of navigating easier, the transponder makes flying safer by allowing the ground-based air traffic controller to see and identify aircraft easily in the control sector. Some avionics equipment is required by law, and in most cases aircraft are outfitted with more avionics than is mandated by law. The equipment required by the FAA depends on the type of flying done by the aircraft, that is, on whether the aircraft is flown for hire, whether it is flown under instrument flight rules, or the class of airspace it is flown in.

For all aircraft, the Federal Aviation Regulations (FARs) require a magnetic compass, a full set of engine instruments, an airspeed indicator, an altimeter, and an emergency locator transmitter. This minimal complement of equipment severely limits the type of flying done by the aircraft, and very few aircraft have such a spartan system. Without radio communications, an aircraft could take off and land only at an airport that does not have a control tower. Additionally, aircraft could not traverse the control zone of an airport that has a control tower. The lack of a communications radio so severely limits the aircraft that only agriculture (crop dusting) and aerobatic aircraft are not fitted with a communications radio, and these aircraft often carry a portable communications set when they are flown any appreciable distance.

Class B airspace requires a VOR or TACAN receiver, a communications radio capable of communicating with the air traffic control center for the terminal control area, and a transponder. Most small aircraft contain this equipment as a minimum. This complement of equipment allows almost limitless daytime VFR operation.

Flight under instrument flight rules requires the same engine instruments as required for VFR, plus the radio communications transceiver, VOR or TACAN navigation receiver, and transponder required for class B airspace flying under VFR. Additional lighting and a spare set of fuses are also required. A rate-of-turn indicator, a slip-skid indicator, an artificial horizon, a directional gyroscope, a sensitive altimeter with barometer adjustment, and a clock capable of displaying hours, minutes, and seconds are required. Flying above 24,000 ft above sea level requires a DME.

Although an ADF is not required by any of the FARs, it is usually installed in small aircraft as an aid in flying into and out of small airports not near a VHF navigation facility.

Aircraft electronics are grouped into four general categories: navigation, communication, surveillance, and control.

Navigation equipment provides information to determine the aircraft's position in one, two, or three dimensions and to provide information for flying the aircraft to a specific location or course. This refers to equipment and systems for en route navigation or for short-range systems used for approach and landing.

Surveillance equipment is used to detect and avoid bad weather, obstacles, or other aircraft. Many surveillance systems require that part of the system be installed in the aircraft while other parts be ground based. The air traffic control radar beacon system, ATCRBS, is a system of this sort. Although the airborne radar transponder is a part of this system, the actual surveillance is done from the ground. The airborne weather radar is a wholly airborne system.

Communications equipment consists mostly of voice radio transmitting and receiving equipment. Some data communications equipment is being installed in aircraft, and this trend will continue. However, the vast majority of aircraft communications takes place by voice.

Aircraft control equipment includes autopilots and other devices that automatically adjust some mechanical system, such as engine synchronizers in multiengined aircraft. With the exception of the autopilot, these systems are specialized.

REGULATION AND STANDARDIZATION

Pilots must be licensed and aircraft certified in all countries throughout the world. Pilots must be examined for their aviation knowledge and demonstrate their flying skills under normal and emergency conditions to a government-appointed examiner. Aircraft must also pass examinations. Like the pilot, an aircraft must demonstrate its ability to fly under both normal and emergency conditions. To ensure ongoing safety, both aircraft and pilots must be periodically reexamined to demonstrate that the abilities originally tested are not lost.

When an aircraft is designed, the design must be tested and analyzed to prove the safety of the design. Once it has been demonstrated to the FAA that the design is inherently safe, the design is awarded a type certificate. This certificate is for a design, not for any hardware. To build aircraft, the type-certified design must be built under approved conditions, and the result is an aircraft, actual hardware, that is type certified.

Type certification is required to obtain an airworthiness certificate, which is needed to obtain registration. The latter allows the aircraft to be flown in the United States.

Once a design has been type certified, no changes may be made to the design without approval. If design changes are made, the type certificate may be modified by a supplementary type certificate, or STC.

Change of design applies not only to the manufacturer, but also to changes made by aircraft owners. As an example, if an aircraft is modified to include new engine types not originally specified on the type certificate, an STC is required. An STC, essentially, is proof that the changes made to the basic design do not compromise the safety of the aircraft. In the case of new engines, engineering calculations and flight tests would be required to prove that the new engine was a safe installation.

Not every aircraft that the new engines are installed in requires the calculations and flight tests. If the tests were performed for an identical aircraft, the data from the previous tests may be used as a basis for an STC. This form of STC is done regularly by manufacturers that provide improved parts for their aircraft; obtaining an STC is a matter of submitting paper work rather than repeating flight tests.

STCs are important because major changes in avionics installations require an STC. Owing to the great advances in electronics technology and the large variety of avionics types, avionics STCs are very common.

GOVERNMENT AGENCIES

The design, licensing, and use of navigation aids are controlled throughout the world by several agencies dedicated to the production of a universal system that will allow international flight with a common set of navigation and communications equipment. These are discussed in their order of authority over matters relating to navigation and other airborne electronic equipment in the United States. Federal agencies in the United States have counterparts in foreign nations, and a similar list of agencies in a foreign land would contain the counterpart agencies in lieu of the U.S. agency. In other respects, the role of regulatory and standardizing agencies is the same throughout the world.

Federal Aviation Administration (FAA).   This organization is the ultimate regulatory agency for flight safety in the United States. The rules and regulations of the FAA, a part of the Department of Transportation of the U.S. government, are contained in the Federal Aviation Regulations (FARs) and cover the rules of flying, requirements for pilots, certification of aircraft, maintenance of airways, and so on. The FARs also provide for punishment in the form of fines and imprisonment if the rules are violated.

Technical standards for avionics are embodied in Technical Standard Orders, or TSOs, which cover not only electronic equipment, but other aircraft systems, such as safety systems, wheels and tires, and so on. When equipment has been tested to TSO standards and has been accepted by the FAA, it is said that the equipment has been TSOed.

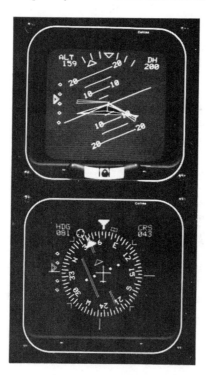

Photo 1-2 With an eye toward improving reliability and versatility, these "indicators" are created with a cathode ray tube display. Photograph courtesy of Rockwell International Corp., Collins Divisions.

Aside from writing and enforcing regulations, the FAA provides the majority of the navigation ground systems, as well as flight assistance and airport tower operations. Research on navigation systems is conducted to improve the effectiveness of electronic navigation aids and to investigate future navigation systems.

The FAA also investigates aircraft accidents and other incidents in cooperation with the National Transportation Safety Board and other agencies. The FAA publishes accident investigation findings and other information for pilots, and offers courses on air safety throughout the United States.

International Civil Aviation Organization (ICAO).   This is an agency of the United Nations headquartered in Montreal, Canada, that defines and publishes standards for navigation procedures and for equipment used in international travel. The policies of ICAO are usually adopted by the major nations of the world as standard. Some navigation aids standardized by ICAO are VHF omnirange, distance-measuring equipment, and the air traffic control radar beacon system. An aircraft equipped with navigation equipment designed to operate with these systems as defined by ICAO can fly into any foreign airport equipped with ICAO-specified ground equipment and can navigate with no degradation in performance. ICAO also specifies the signal formats, equipment performance levels, and assigned frequencies for navigation aids.

Federal Communications Commission (FCC).   The Federal Communications Commission is the regulatory agency for all radio-transmitting equipment in the United States. All airborne navigation and communication equipment that transmits a radio signal must be licensed to the equipment owner by the FCC. This equipment includes communications transceivers, radar transponders, radar altimeters, and distance-measuring equipment.

Repair technicians must be licensed to repair radio transmitters. Radio repair shops must be certified by the FAA, but the technicians must be licensed by the FCC. A general class radiotelephone license is required for anyone responsible for the repair of any radio transmitter. Infractions of FCC rules can bring fines and imprisonment.

International Telecommunications Union (ITU).   Headquartered in Geneva, Switzerland, the ITU is an international organization for radio-frequency spectrum utilization. The organization represents practically every sovereign nation in the world and assigns sections of the radio spectrum to navigation services. At ITU's World Administrative Radio Conference, or WARC, member countries vote on the disposition of the radio-frequency spectrum. The ITU has no police powers, and each individual country must have its own regulatory agency to legislate ITU decisions into law and must enforce that law.

RTCA, Inc., formerly the Radio Technical Commission for Aeronautics. RTCA, headquartered in Washington, D.C., is comprised of government and industry interest groups. Both manufacturers and avionics users are members that represent industry. Government interest groups include the FAA and military users. Civilian users are represented by various pilot's associations, airlines, and other groups.

The purpose of the RTCA is to define and set forth standards for avionics equipment. The standards are then published and made available to the public and government agencies for comments.

The subject matter to be considered by the RTCA is at the suggestion of industry or the government. Committee meetings are held over a period of time while the specifications for a particular navigation system are discussed.

Although the RTCA has no official power, the FAA often accepts the specifications outlined by the RTCA and makes them law by declaring them to be the basis for a TSO. The FAA can declare the entire RTCA document as the TSO or can add to or delete from the RTCA document.

Many foreign governments look to the United States for model avionics regulations. Because of this, the RTCA works closely with many foreign governments during the development phase of a specification, and particularly with the European Organization for Civil Aviation Electronics, or EUROCAE.

Aeronautical Radio Incorporated (ARINC).   ARINC is a nonprofit corporation supported by revenues received for communications services supplied. ARINC was formed and is supported by the scheduled airlines in North and South America and provides communications networks, both radio and telephone, for the airlines in the United States. ARINC has standardized the interconnection systems used in the avionics of large aircraft so that equipment of several manufacturers can be interchanged. The size of the equipment and the location and type of connectors of avionics equipment are defined so that the units will interchange electrically. In addition, performance is specified for various operating conditions. ARINC standards have been adopted not only by airlines, but by many manufacturers of smaller aircraft.

Government Agency Approval

In the United States, aircraft electronics equipment is approved by the FAA or the FCC or both. All equipment that transmits a radio signal, such as a communications transceiver, DME, radar transponder, radar altimeter, or weather radar, must be type accepted by the FCC and be covered under a radio station license.

Type acceptance must be obtained by the manufacturer before the transmitting device can be used under the authority of a radio station license. To obtain a type acceptance, the manufacturer must submit the equipment to a series of tests and provide the results to the FCC. In addition, schematics, photographs, and descriptions of the circuits must be submitted to the commission for approval. If the equipment passes the required tests and the data are accepted by the commission, a type acceptance is granted, which allows the equipment to be sold and licensed in the United States.

All receiving equipment operating below 900 MHz must also be approved by the FCC, even if the equipment is a part of a transmitter that must be type accepted. Therefore, VHF navigation receivers and ADFs must be type accepted by the FCC before they may be sold.

The FAA approves avionics equipment by granting a TSO to the avionics system (see Table 1-1). To obtain a TSO, the manufacturer must submit the equipment to testing and

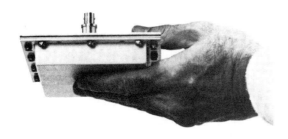

Photo 1-3 Modern solid-state technology has reduced the radar altimeter to only two assemblies: an antenna and an indicator. Photograph courtesy of Bonzer, Inc.

provide the data of the test to the FAA for approval. In addition, the FAA will assess the manufacturer's quality control procedures to ensure that the equipment granted the TSO will continue to provide the level of performance expected of TSOed equipment. TSOed avionics equipment is not required for flight under the general flight rules, part 91 of the FARs, except for flights into class B airspace. In these areas, a TSOed transponder is required. The TSO is a guarantee to the customer of a specific level of performance monitored by the FAA.

The most comprehensive set of specifications for the performance requirements of civil aviation electronics systems was created and distributed by RTCA. Some earlier documents are famous because they set forth the principles of a system that became the standard for the world. Most notable are the VOR and DME systems. Since the late 1940s, RTCA has updated distributed documents describing the required minimum performance specifications for airborne electronic systems. Many FAA TSOs call out an RTCA document as the required operational specifications for a particular avionics system. Some TSOs, however, do not refer to an RTCA document, but include in the FAA orders the complete text of the operational specifications. The trend in the FAA is toward generation, maintenance, and distribution of the minimum operational specifications to the aircraft industry by RTCA. When an RTCA document is lacking but there is an existing TSO, the RTCA has plans in most cases to create an applicable document in the future. Table 1-1 lists FAA TSO numbers and the RTCA documents that provide the basis for operational performance. Because the TSOs are periodically changed and additional TSOs added, the table is not representative of the present state of law. Several RTCA documents are different from the operational performance standards that affect only one type of avionics system. These are general documents, such as environmental requirements and software documentation requirements.

TABLE 1-1

TSO number	System	RTCA document
C31C	High-frequency transmitting equipment	None
C32C	High-frequency receiving equipment	None
C34C	Glide slope	D0132
C35D	Marker beacon	D0143
C36C	Localizer	D0131
C37C	VHF communications transmitting equipment	D0157
C38C	VHF communications receiving equipment	D0156
C40B	VOR receiving equipment	D0153A
C41C	ADF receivers	D0142
C50B	Audio interphone	None
C57	Headsets and speakers	D090
C58	Microphones	D091
C60A	LORAN-C receiving equipment	D0159
C63B	Weather radar	D0134
C65	Doppler radar	D098
C66B	Distance-measuring equipment (DME)	D0151A
C67	Radar altimeter	D0103
C68	Automatic dead-reckoning equipment	D0104
C71	Dc-to-dc converter	D060
C73	Static power converter	None
C74	ATC transponder	D0144
C87	Low-range radar altimeter	None
C94	Omega receiving equipment	D0164A
C104	Microwave landing system receivers	D0177
C118A	TCAS I, traffic alert and collision avoidance system	None
C119A	TCAS-II, traffic-alert and collision avoidance system	D0185

RTCA environmental documents are used for most systems appearing in the TSO list. The older DO-138 was in use for many years and has been supplanted in later TSOs by DO-160C, which is the current document. This document has several categories for most environmental tests. For example, several categories of vibration are included in the document to correspond to the various types of vibration found in different aircraft types. More low-frequency vibration testing is performed for helicopters, which experience more low-frequency vibration due to the rotor blades. On the other hand, higher-frequency vibration testing is performed for equipment being installed in jet-powered aircraft. DO-160 has been changed a number of times to reflect the evolution of avionics. Most recently, DO-160C has been updated to include tests for high-intensity radiated fields, or HIRF. Because electronics have become increasingly critical on aircraft and the microprocessor is involved, upsetting a microprocessor has become a dangerous event. Recognizing this, DO-160C includes new tests involving single, energetic pulses of electromagnetic fields.

Practically all the environmental categories can be tailored to meet the aircraft type. A sequence of letters is used to describe the environmental categories. The environmental category is displayed on the TSOed equipment on an identification tag. The 14 environmental categories are outlined in Table 1-2. Each letter or number in the environmental identification represents the category of the test. The first number or letter in the sequence

Photo 1-4   A flight management system that stores all pertinent information about a flight and controls several navigation systems. Photograph courtesy of Rockwell International Corp., Collins Divisions.

TABLE 1-2   ENVIRONMENTAL CATEGORIES

1.	Temperature and altitude (two letters)
2.	Humidity test
3.	Vibration test
4.	Explosion test
5.	Waterproofness test
6.	Hydraulic fluid test
7.	Sand and dust test
8.	Fungus resistance test
9.	Salt spray test
10.	Magnetic effect test
11.	Power input test
12.	Voltage spikes test
13.	Audio-frequency conducted susceptibility test
14.	Electromagnetic compatibility test

represents the temperature and altitude testing, and the second number or letter represents the category of the humidity tests. The third number or letter represents the vibration test category, and so on. Typical environmental category designators would be as follows:

DO-160C Env. Cat. A2AKXXXXXXAAAAA

In some cases, the manufacturer will test the equipment to more than one category, and in this case an environmental category might look like this:

DO-160C Env. Cat. /A1D2/A/MNO/XXXXXXAAAAA

In this example, the equipment has been tested in categories A1 and D2 for altitude and temperature. Vibration has been tested in categories M, N, and O. Category X, in many but not all cases, indicates that the equipment has not been tested at all for that category. For a complete description of the testing performed, the RTCA document DO-160C should be consulted.

Many airborne navigation systems require radio-wave propagation, and thus aircraft carry both radio transmitters and receivers. The frequencies used by airborne navigation and communications systems span from very low frequencies to the microwave region. The range of frequencies used for radio are called *radio frequencies,* or RF. These frequencies are divided into *bands* as shown in Table 1-3. Long-distance navigation systems occupy the lowest frequencies, below 15 kilo hertz (kHz), as shown in Fig. 1-1. Nondirectional beacons occupy the region from about 200 kHz to the bottom of the standard broadcast band. Because of the similar radio propagation of the standard broadcast system, radio broadcast transmitters may be used for direction finding. The now obsolete LORAN-A long-range navigation system occupied a few frequencies above the standard broadcast band around 2 MHz. The entire high-frequency portion of the spectrum is used for long-distance communications on specific assigned frequencies. The marker beacon transmitter occupies a single channel at 75 MHz at the lower end of the VHF spectrum, while a very broad band of frequencies for navigation and communications is assigned to aircraft use between 108 MHz (just above the standard FM broadcast band) and 136 MHz. The glide slope portion of the instrument landing system is assigned a block of channels in the 300-MHz region. The frequency range from 978 to 1213 MHz is reserved for distance-measuring equipment and air traffic control secondary radar transponders. Radar altimeters occupy a band of frequencies in the C band around 4.2 gigahertz (GHz), while microwave landing systems, MLS, occupy other frequencies in the C band between 5.03 to 5.09 GHz. The highest frequency currently in use for airborne avionics use is the X band, at about 10 GHz, where airborne weather radar systems operate.

TABLE 1-3

Frequency range	Nomenclature	Abbreviation
3 kHz–30 kHz	Very low frequency	VLF
30 kHz–300 kHz	Low frequency	LF
300 kHz–3 MHz	Medium frequency	MF
3 MHz–30 MHz	High frequency	HF
30 MHz–300 MHz	Very high frequency	VHF
300 MHz–3 GHz	Ultra high frequency	UHF
3 GHz–30 GHz	Extreme high frequency	EHF
30 GHz–300 GHz	Super high frequency	SHF

A graphical representation of the electromagnetic spectrum showing the location of the aviation frequencies is shown in Fig. 1-1. Aviation-use frequencies are spread throughout the entire frequency spectrum and were chosen for a required characteristic of the frequency. The very low frequencies are required for long-distance navigation because of their predictable propagation characteristics. The low frequencies of the nondirectional beacons were chosen for their predictable propagation characteristics at short distances. The frequencies in the VHF and UHF portion of the spectrum were chosen for their line-of-sight

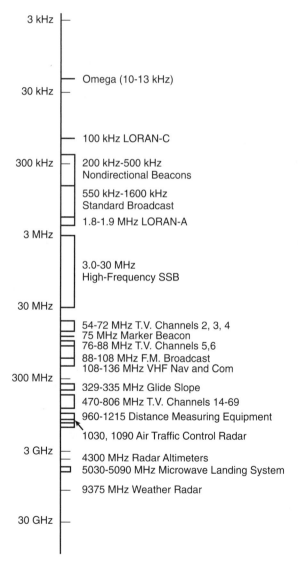

Figure 1-1   Electromagnetic radio spectrum showing aviation services and some broadcast services.

propagation and relative freedom from atmospheric noise. The very high microwave frequencies used by weather radar at the top of the electromagnetic spectrum are necessary because the very short wavelengths are easily reflected from raindrops in storm clouds.

AIRCRAFT CHARACTERISTICS AND NOMENCLATURE

To the student of aviation electronics, it is helpful to have a knowledge of the basic components of aircraft. There are three general classifications of aircraft: lighter than air, rotorcraft, and airplanes. By far the most common are airplanes, and for this text only airplanes will be considered. Dirigibles and helicopters, the most common forms of lighter-than-air craft and rotorcraft, are a part of the U.S. fleet and do contain avionics, but represent a small portion of that fleet.

Airplanes are further divided into two categories: land and sea. This text will focus on the land airplane, which represents the vast majority of registered aircraft.

Referring to Fig. 1-2, the airplane consists largely of a *fuselage,* which is the passenger and cargo part of the aircraft and flying surfaces. A *wing* provides the necessary lift for the airplane, but also contains various systems required for the operation of the aircraft. The most common payload contained in the wing is fuel, although motors, landing lights, and other items are also found within the wing.

There are three axes of the airplane, just as there are three orthogonal axes. The *roll axis* is around a line from the front to the rear of the aircraft. The *pitch axis* is oriented around a line that runs through the center of the wing. The *yaw axis* is around the vertical axis of the airplane.

The wing also contains two movable flying surfaces, the *ailerons* and the *flaps*. The flaps alter the shape of the wing, which affects the lift and assists in takeoffs and landings. The ailerons are used to control the airplane around the roll axis. The ailerons move in opposite directions. When one aileron is in the up position, the other is in the down position, which produces different lift for each wing and causes the aircraft to roll.

Two flying surfaces are at the rear of the aircraft, the *horizontal* and *vertical stabilizers.* These surfaces do not provide any significant lift but are used to provide stability and control of the pitch and yaw axes. Included as a part of the vertical stabilizer is the *rudder,* which provides rotation around the yaw axis. Included as a part of the horizontal stabilizer is the *elevator,* which provides rotation around the pitch axis. In some aircraft designs, the pitch and yaw controlling flying surfaces are mounted in the front of the aircraft, but these are rare.

Aircraft propulsion is provided by either jet engines or propellers. Large aircraft almost universally are propelled by jet engines, whereas most smaller aircraft are propelled by propellers.

Most aircraft are constructed of sheet aluminum riveted to an internal frame. The actual strength of the aircraft comes from the skin of the aircraft. Early aircraft designs used various forms of frames covered with thin material, but the trend toward obtaining the aircraft's strength from the skin is nearly universal.

Aluminum is a reasonably good conductor of electricity and is used to provide a path for current flow for some aircraft electronics devices. The fuselage contains a number of aircraft radio antennas, and the conducting fuselage is needed for the correct operation of

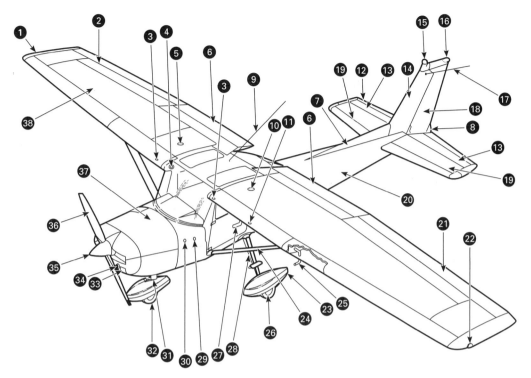

1.	Right wing position light (green)	13.	Left and right elevators*	24.	Wing strut
2.	Right aileron	14.	Vertical stabilizer* (fin)	25.	Fuel vent
3.	Fresh air vent			26.	Main landing gear
4.	Outside air temperature guage	15.	Flashing beacon	27.	Pitot tube
		16.	Tail position light (white)	28.	Main landing gear strut
5.	Right fuel tank cap	17.	VOR antenna	29.	Radio cooling vent
6.	Wing flaps	18.	Rudder*	30.	Static air vent
7.	Dorsal fin	19.	Left and right horizontal stabilzers*	31.	Oleo strut
8.	Rudder trim tab			32.	Nose landing gear
9.	Communications antenna	20.	Fuselage	33.	Carburetor air intake
		21.	Left aileron	34.	Taxi and landing light
10.	Left fuel tank cap	22.	Left position light (red)	35.	Propeller spinner
11.	Stall warning device			36.	Propeller
12.	Elevator trim tab	23.	Wheel fairings	37.	Engine cowling
				38.	Wing

*Components of empennage

Figure 1-2   Nomenclature of an airplane.

the antenna. In larger aircraft, a nonconducting dome is provided at the nose of the aircraft behind which are mounted several antennas that need to be shielded from the environment. Typically, a weather radar antenna and a glide slope antenna are mounted behind this dome, which is called a *radome*.

Some parts of an aircraft are made from composite materials. These materials are usually strong filaments of various types of materials, such as glass fibers and Kevlar. These composites produce a very strong structure with light weight. Like the radome, these materials do not conduct electrical current and cannot be used with antennas for which a metallic fuselage is required.

Most aircraft used for revenue flight are pressurized. This technique uses an air compressor to increase the air pressure within the cabin of the aircraft above the outside ambient. There is insufficient air pressure to breathe above about 10,000 feet. Not all of the aircraft is pressurized. Because additional strength is required where there is a pressure difference, only the areas occupied by passengers and some baggage are pressurized. Thus some areas of the aircraft are not pressurized yet may contain various equipment. The same is true of heating.

Because of the potential for air leaks, the walls that separate the pressurized areas from the nonpressurized areas are called *pressure bulkheads.* Because avionics equipment can be placed outside the pressurized part of the aircraft, it is important when wires pass through the bulkhead that either airtight connectors be used or that the wires passing through the bulkhead be sealed against air leakage.

REVIEW QUESTIONS

1.1. What were the three instrumentation requirements for early zero-visibility flight?

1.2. What was the four-course radio range?

1.3. What are some problems of low-frequency navigation systems?

1.4. Define DME, RNAV, NAV radio, COM radio, and class B airspace.

1.5. What is the purpose of the Federal Aviation Administration?

1.6. What is the scope and purpose of FARs and TSOs?

1.7. What is a type certificate?

1.8. What is an STC?

1.9. What role does the FCC play in aircraft electronics?

1.10. How do ICAO recommendations become law?

2

A Review
of Electrical Circuits

CHAPTER OBJECTIVES

This chapter will review basic circuits and electrical components. Radio-wave propagation is also covered.

After studying the material contained in this chapter the student will understand Ohm's law; Kirchhoff's law; the characteristics and types of resistors, capacitors, and inductors; resistors and capacitors in series and parallel; diodes, transistors; and integrated circuits. The student should also understand basic microprocessor fundamentals and radio-wave characteristics and propagation.

The student will be able to calculate power, current, and voltage for ac and dc circuits and express powers in dBm. The use of the decibel notation will be understood.

The student will be familiar with the components and functions of the elements of the microprocessor system.

INTRODUCTION

Electricity has always been a part of aviation from the very first days of powered flight. Even though the Wright flyer had only magnetos to spark the engine, motors, lights, and simple instrumentation began to appear in aircraft as early refinements. After the demonstration of blind flight in 1929, radio equipment began to proliferate until the present, when an aircraft contains electronic systems that would transcend the wildest imaginations of the early pioneers.

It is necessary to understand the basics of electrical circuits to fully comprehend aircraft electronic systems. Electricity is the flow of electrical charges through conductors and electrical circuit elements such that it performs useful functions. These functions are diverse—lighting annunciators, turning motors, moving indicators, performing logical decisions, communicating, and on and on.

Before we can fully appreciate charges, conductors, and electricity, we must have an understanding of the nature of matter. Electrical charges are a part of all matter. All matter consists of molecules, which are composed of atoms. *Molecules* are the smallest part that a substance can be divided into and still retain the characteristics of that substance.

As an example, water consists of a molecule that has two hydrogen atoms and one oxygen atom for each molecule. Although it would be a very small amount of water, one molecule of water would have the same characteristics as a gallon of water. The same characteristics means that it would react chemically the same, have the same thermal characteristics, the same density, and so on.

If the water molecule is broken apart into two atoms of hydrogen and one atom of oxygen, the characteristics of water are lost. Hydrogen and oxygen are gases, they react entirely differently than water, their specific gravities are different, and so on.

It is easy to break molecules apart into their constituent atoms, but it is very difficult to break atoms apart into different substances. Powerful nuclear particle accelerators can be used to break atoms apart, but normal chemical and physical reactions are not capable of splitting the atom; thus, for practical purposes, the atom represents the smallest division of matter.

The atom has constituent parts; it consists of a *nucleus* and *electrons.* The nucleus consists of *protons* and *neutrons.* The number of protons in the nucleus is called the atomic number and determines the element. As an example, the atom with six protons is carbon, while nitrogen has seven protons. Certainly, carbon and nitrogen are very different, yet they are close in atomic number.

There are at least as many neutrons as protons in the nucleus and usually a few more neutrons. Atoms with different numbers of neutrons in the nucleus are called isotopes. In nature, it is not unusual for a number of different isotopes of common elements to exist. The chemical characteristics for the different isotopes are identical because it is the number of protons that changes the characteristic of the atom. However, different isotopes do behave differently when nuclear reactions are involved.

The three basic subatomic particles, protons, neutrons, and electrons, have simple characteristics of mass, size, and charge. The electron is the smallest and lightest of the three, while protons and neutrons are much larger and massive.

A very important characteristic of two of the three subatomic particles is *charge.* The electron is negatively charged, while the proton has exactly the same amount of charge but positive. The neutron has no charge. An atom that consists of equal numbers of protons and electrons will have a net zero charge. If one electron is removed from the atom, the net charge is positive because there are more protons than electrons. Atoms with unequal amounts of protons and electrons are called *ions.* Chemical reactions involve the interplay of electrons, and elements that have a net charge behave very differently from neutral ions in chemical reactions.

One model of an atom, which would not suit all purposes but will suffice here, is a nucleus with electrons orbiting the nucleus, much like the planets orbiting the sun. The negatively charged electrons are bound to the nucleus by the attraction of the positively charged nucleus. The electrons are not tightly bound and can be forced out of their orbits by the application of small amounts of energy. Once an electron has been forced from its orbit, it is free to roam until it finds another atom to orbit.

Electricity is the flow of charges, and it can now be seen that electrons are a good choice for electrical charge because they are mobile. Some materials have more mobile electrons than others and are called *conductors.* Materials in which the electrons are not available for conducting electricity are called *insulators.* Good conductors are often metals, such as copper, silver, and gold; a number of materials such as glass, ceramics, and plastics tend to be poor conductors.

Electrical current conduction is not a mass of electrons streaming through space from one end of a wire to another. In metallic conduction, an electron is pulled from its nucleus due to the electrical potential applied to the conductor and drifts only a short distance to a nearby atom that has lost an electron and has a net positive charge, which attracts the free electron, which is eventually captured and orbits the nearby nucleus. Rather than streaming from one end to the other of a conductor, the electrons move only a short distance from one atom to a nearby neighbor. This is very much like water in a pipe. If a gallon of water is poured into a long pipe already filled with water, a gallon exits from the other end, but it is not the same gallon that was poured into the pipe. All the water in the pipe moved a short distance along the pipe so that one gallon poured from the far end.

An interesting quirk of history should be mentioned at this time. Electrical circuits were investigated and understood before the electron was discovered. Having no knowledge of the atom and its constituent parts, the early electrical researchers assumed that the charges were positive and analyzed circuits in terms of *current flow,* which was the movement of positive charges. Early researchers, such as Benjamin Franklin, knew there were two types of electricity, but there was no identification of the source of these charges. The electron was discovered in 1897 by the scientist J. J. Thompson, but 15 years prior, Thomas Edison had built a power generating station in New York City, electrified street cars were operating, electric railroads had run for more than 50 years, and a new profession was being taught in colleges—electrical engineering. What a surprise it must have been when physicists realized that the actual charge flow was electrons that moved in the opposite direction from the hypothesized current. This discovery did not negate any of the circuit analysis based on current flow, because negative charges moving from point *A* to *B* are exactly the same as positive charges moving from point *B* to *A,* in spite of the fact that no positive charges are in motion.

It does not matter whether circuits are analyzed using electron flow or current flow. However, once one convention is accepted, all circuit analysis must be done with the same convention. It is not permitted to change conventions midway through an analysis. This textbook will use the original convention of current flow, which is positive charge motion and is characterized as current flow from a higher electrical potential to a lower potential.

Just what makes current flow? First, there must be a reason for the charges to move, which is provided by a difference in electrical potential energy called *voltage.* Because electrons are charged particles, they are attracted to other charged particles. The attraction of a charged particle by the presence of another charged particle is what causes current flow. Second, there must be a path for current to flow from one electrical potential to another.

Conductors provide easy paths for current flow. A conductor is a material in which a small amount of energy is required to pull an electron from its nucleus. A number of common elements make excellent conductors, such as copper, silver, and gold.

The opposite of a conductor is an insulator which does not support current flow. In an insulator, the electrons are difficult to remove and do not provide current flow. However, if enough voltage is applied to an insulator, the electrons will be dislodged and become available for conduction; when this occurs, the insulator is said to "break down." When this happens, a lot of energy is dissipated and the material can be damaged. A very common example of breakdown is lightning. Air is normally an insulator, but when the electrical potential of clouds builds to enormous amounts, the air breaks down and provides a current path for the discharge.

Interestingly, most good conductors of electricity are also good conductors of heat, and most insulators of electricity are also good insulators of heat. This is actually a problem, because there are many situations for which a thermal path is desired, such as from a hot transistor to a chassis or heat radiator, and an electrical path would cause a short circuit. Sophisticated compounds have been developed that have good conductivity to heat but remain electrical insulators.

How does a current flow? A *battery,* whose schematic symbol is shown in Fig. 2-1(a), provides a source of electrical potential energy difference. A *schematic* is a circuit drawing. Just as a mechanical drawing shows the mechanical arrangement of some device, a schematic shows the electrical arrangement of an electronic circuit. The positive terminal of the battery is at a higher potential energy than the negative. However, a battery with no electrical connections does not provide current flow.

A wire, which is a current path, conducts no current sitting on a table. The wire is a conductor because the electrons in the wire are free to move; they are not tightly bound to their atoms. There must be a difference in electrical potential from one end of the conductor to the other to give the charges in the wire a reason to move.

If the wire were connected from one terminal of the battery to the other, we have the ingredients for current flow—a source of potential difference and a conductor that provides a current path. From a practical standpoint, such a circuit would produce a warm piece of wire and a dead battery after a short time because there is nothing in the circuit to limit the amount of current.

One electrical element that limits the amount of current flow is the *resistor.* The schematic symbol for a resistor is shown in Fig. 2-1(b). Wires are shown as lines.

Let us construct a circuit using a battery and a resistor, connected with wires, as shown in Fig. 2-2. The resistor limits the current in an electrical circuit and follows this simple relationship:

$$I = \frac{E}{R} \qquad\qquad (2\text{-}1)$$

This is *Ohm's law* and it is one of the foundations of the electrical art. Let us apply this simple formula to the circuit of Fig. 2-2 to calculate the circuit current.

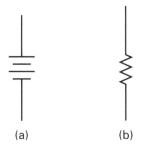

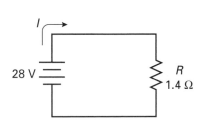

(a)   (b)

Figure 2-1 Schematic representation of (a) voltage source
and (b) a resistance.

Figure 2-2 Simple electrical circuit
showing a voltage source and a resistance.

The battery is 28 volts (V) and the resistor is 1.4 ohms (Ω), and thus the current is 20 amperes (A). Remembering the current flow convention, current flows from a higher potential to a lower potential; the polarity of the voltage drop across a resistor is positive where the current enters the resistor.

Another circuit parameter needs investigation. The battery has a specific voltage, which was defined as a difference in electrical potential energy. When charges are permitted to flow from one electrical potential energy to another, energy is transferred from the voltage source to something else. Each charge transfers a specific amount of energy, which is proportional to the potential difference. Thus, the amount of energy transferred per second is proportional to the number of charges that flow in 1 second (s), which is a measure of current and the potential difference. Mathematically, this is expressed as

$$\text{energy transfer per second} = kEI \qquad (2\text{-}2)$$

where k is the constant of proportionality. The rate of transfer of energy per unit of time is *power*. Wouldn't it be nice if the equation was arranged so that the constant, k, could be 1? This is exactly what is done by using the definition of current as 1 A, which is equal to 1 coulomb (C) of charge per second. The charge of 1 C is equal to the charge of 6.25×10^{18} electrons, which is the key to Eq. (2-3) having a constant of 1. The energy is 1 joule (J) per coulomb and the energy transfer is in watts. The equation is now

$$P = EI \qquad (2\text{-}3)$$

where the unit for P, power, is watts.

Let us return to the previous example of a 1.4-Ω resistor and a 28-V battery for which the current was calculated as 20 A. The power delivered from the battery to the resistor is

$$P = EI = 28 \times 20 = 560 \text{ W} \qquad (2\text{-}4)$$

This problem started with a battery and a resistor and calculated the current using Ohm's law and then used the current to calculate the power delivered to the resistor. Ohm's law can be used with the power equation to derive two other power relationships.

$$P = \frac{E^2}{R} \quad \text{and} \quad P = I^2R \qquad (2\text{-}5)$$

These two equations were derived by using Ohm's law to replace I with E/R for the first equation and to replace E with IR in the second equation.

These equations may be used to calculate all the circuit parameters of a simple circuit with one resistance. Certainly, electrical circuits are much more complex than this, and a few additional laws are required. These additional laws will allow us to analyze circuits with more than one voltage source and more than one resistor.

The first law is called Kirchhoff's current law, which deals with a junction. In this case more than one conductor is connected together as shown in Fig. 2-3. Kirchhoff's current law says: *The sum of all currents into a junction is equal to all currents leaving a junction.* This relationship is obvious. Since electrical current is the actual flow of physical charges, if the current into a junction does not equal the current leaving that junction, there must be an accumulation of charges at the junction. It is known that this does not happen. Conductors do not have the ability to store charges.

If we adopt the convention that current that flows into a junction is considered positive current and current leaving the junction is negative current, then Kirchhoff's current law could be rewritten as follows: *The sum of all the currents into a junction is zero.*

Although this current law is a powerful tool to analyze circuits, a second Kirchhoff's law is required, which states: *The sum of all the voltages in a complete circuit is equal to zero.*

This law is not as obvious as Kirchhoff's current law, but can be proved rather simply. Consider Fig. 2-4. The current through each resistor is the same because of Kirchhoff's current law. The voltage of the battery is clear. By Ohm's law, each resistor has a voltage associated with it given by $E = IR$. According to Kirchhoff's voltage law, the battery voltage equals the sum of all the voltage drops across each resistor. This can be proved by a simple law of physics, that of the conservation of energy. The energy provided by the battery is EI joules every second, where E is the battery voltage and I is the circuit current. All the energy supplied by the battery is transferred to the resistors. The sum of the energy dissipated by the resistors each second must be equal to the energy supplied by the battery. The energy dissipated each second is the power dissipated, and the following equation can be written:

$$P = EI = E_1 I + E_2 I + E_3 I = I(E_1 + E_2 + E_3)$$

$$E = E_1 + E_2 + E_3$$

Therefore, the voltage across the resistors must equal the battery voltage. This law can be expanded to any circuit that contains energy sources and resistors.

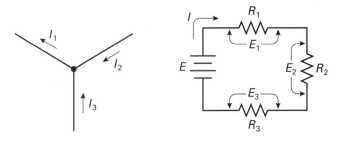

Figure 2-3   Electrical junction.   **Figure 2-4**   Series circuit.

Resistors in Series and Parallel

In a series circuit, shown in Fig. 2-4, there is only one current path, and thus the same current flows through every part of the circuit. A break in the circuit anywhere will cause the cessation of current.

In a parallel circuit, there is more than one current path. Thus, if a part of the circuit is broken, there is still some current. But the currents within the circuit are not necessarily the same. As in the previous example for the series circuit, these are three different ways of saying the same thing. An example of a parallel circuit is shown in Fig. 2-5.

Series and parallel circuits are simple and are characterized by simple definitions. Most circuits consist of much more complex circuits and are not strictly series or parallel. However, most circuits can be divided into very simple series and parallel circuits that may be analyzed with easy calculations and assembled for an overall solution.

Some general rules concerning resistors in series and parallel can be derived using Ohm's and Kirchhoff's laws. Refer to Fig. 2-4. What value of resistor would be required to replace the three individual resistors with one resistor? Using the conservation of energy, the total energy dissipated by the one equivalent resistor must be the same as the sum of energies dissipated by the three resistors when connected to the same battery. Therefore,

$$P = I^2 R_1 + I^2 R_2 + I^2 R_3 = I^2 R_{eq} \tag{2-6}$$

where R_{eq} is the equivalent resistance. By dividing all terms of the equation by I^2, we arrive at the result that $R_{eq} = R_1 + R_2 + R_3$. It is clear that the results would be similar for four or five or any number of resistors.

To find the equivalent resistor for a number of resistors in parallel, as shown in Fig. 2-5, the same conservation of energy can be used. The criterion is exactly the same as for the series resistors: the sum of energy dissipated for the three resistors must be the same as would be dissipated from a single equivalent resistor. This relationship is stated as

$$P = \frac{E^2}{R_1} + \frac{E^2}{R_2} + \frac{E^2}{R_3} = \frac{E^2}{R_{eq}} \tag{2-7}$$

Dividing the equation by E^2 results in the following relationship between the equivalent resistor and the individual resistance values.

$$R_{eq} = \frac{1}{\dfrac{1}{R_1} + \dfrac{1}{R_2} + \dfrac{1}{R_3}} \tag{2-8}$$

Resistors convert electrical energy to heat energy, which is not very useful unless there is a need to heat the interior of electronic equipment. Almost every electronic circuit makes

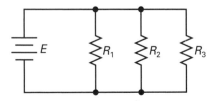

Figure 2-5   Resistors in parallel.

use of resistors for a number of reasons other than heating the interior. The heating effect is an undesirable side effect. One application of resistances is to reduce signal levels. As an example, some circuits require voltages that are exactly one-half of the power supply voltage. The circuit shown in Fig. 2-6 shows a circuit using two resistors that accomplishes this task. The power supply voltage is shown as a battery, and two resistors of any value are used. The voltage across R_2 as a function of the supply voltage can be written with the aid of Ohm's law. The voltage drop across R_2 is the value of R_2 times the current through the resistor. The current through R_2 is the same as anywhere else in the circuit, which is the supply voltage divided by the equivalent resistance, or $I = E/(R_1 + R_2)$. Therefore, the voltage across R_2 is

$$V_2 = E \frac{R_2}{R_1 + R_2} \tag{2-9}$$

If R_1 and R_2 are the same value, the voltage is always one-half the supply voltage; but any fraction of E could be set by adjusting the values of R_1 and R_2. This circuit is called a *voltage divider,* and the resistors are chosen to have as high a value as the circuit will allow to reduce the amount of heat generated.

Resistors come in all sizes and styles depending on the application. Values span from a few thousandths of an ohm to millions of ohms. The physical size of a resistor attests to the resistor's ability to dissipate power. The larger resistors are capable of dissipating more power. Very high power resistors often have provisions for mounting to a heat sink to aid in dissipating the power.

Another important characteristic of a resistor is the tolerance of the resistance. Most common resistors have a tolerance of about 5%, which is suitable for most circuits. In precision measurements, resistor tolerances of 1% and better are used. Very special resistors can be made with tolerances of 0.001% for very critical applications.

Resistors are made of a number of materials. High power resistors are usually made from a high-resistance wire. The metallic wire allows high temperatures to be tolerated without destroying the resistor or causing a large change of resistance. Carbon is used in many resistors either as a block of carbon or a thin carbon film on a ceramic substrate. For precision resistors, thin metal films are used.

Switches and Circuit Breakers

Some simple and important circuit elements are switches and circuit breakers. A large number of circuit breakers are used in aircraft systems so that failed equipment will not cause widespread damage. Many modern circuit breakers are a combination of switch and circuit breaker that reduces the number of controls that have to be installed in an aircraft.

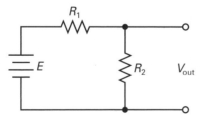

Figure 2-6 Resistive voltage divider.

An *electrical switch* is a mechanical device that allows the current through a circuit to be interrupted. Two metallic conductors are brought together to permit current flow and are separated to interrupt the current flow. A form of actuator is provided, which could be a short handle, a rocker-type actuator, or a push switch. Not all switches are manually operated. Very important applications of switches are to determine door and hatch status and to sense landing gear position.

A *circuit breaker* is also a switch except the switch can be turned off electrically. A common form of circuit breaker uses an electromagnet to electrically operate a switch. The current for the electromagnet is the actual circuit current. The switch and electromagnet are designed such that a specific amount of current is required to overcome the mechanical resistances and activate the switch, which is called *tripping* a breaker. Most circuit breakers can be operated manually as a switch, as well as automatically.

An alternative method of protecting circuits from overload is the *fuse.* This device is a thin conductor that will melt and cause an interruption of current if the current should exceed a design maximum. Unlike circuit breakers, once a fuse has opened due to excessive current, it must be replaced before the circuit can be reenergized. Fuses are usually employed when it is not desirable to attempt reenergizing and the equipment should be investigated by a repair technician before being placed back into service.

Two classes of fuses are in common use. The conventional fuse, often called *fast acting,* is capable of interrupting current in a minimal amount of time. In some applications where short bursts of current are common, such as the starting current for a motor, a slow-acting fuse is used to prevent unnecessary fuse failures.

A magnetically operated switch is the *relay.* The relay has an electromagnet that pulls the movable part of a switch. The purpose of a relay is to allow remote switching of a circuit. An application of the relay would be to switch on a high-power generator using a small switch on the instrument panel of the aircraft. If this were done with a conventional switch, huge wires would have to be run from the generator to the instrument panel, where a huge switch would be mounted to accomplish the task. When a relay is used to perform the switching function, only enough energy to activate the relay is required to be switched from the instrument panel, which usually involves rather small wires and switches.

Relays are available in an entire spectrum of sizes, from very small relays used to switch very sensitive signals to massive relays that perform such tasks as providing the current necessary to turn an engine starter motor. These large relays are also called *contactors,* and a 1000-A contactor is common.

The discussion of electrical fundamentals has been, to this point, about direct current, that is, voltages and currents that are provided by batteries so that the direction and magnitude of flow never changes. There are very important needs for currents that change direction and magnitude in specific ways. Half of the story of electricity is *alternating current.* Around the turn of the twentieth century, two giants of the electrical industry, Edison and Tesla, argued about the advantages of direct current and alternating current, respectively, for electrifying the world. Today their argument would sound foolish; both kinds of electrical current are of extreme importance.

Before we can investigate alternating current we must understand just how this current alternates. The most common kind of alternating current has a waveform that varies in time as the sine of an angle, as shown in Fig. 2-7.

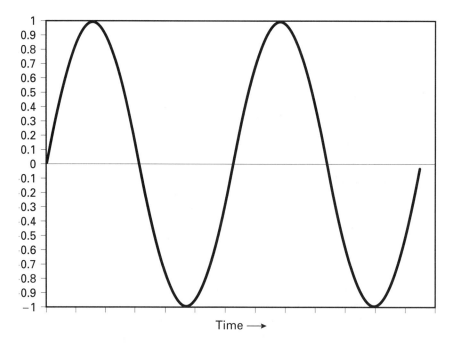

Figure 2-7   Common form of alternating current.

What makes the sine function so common in alternating current? The sine function is the link between the mechanical motion of a generator and the resulting output voltage. But there are many more wonderful characteristics of the sine function that make that waveform valuable in electrical circuits. Because so many electrical circuits generate and use sine waves, this is one of the predominant waveforms found in electronics.

The sine function would be generated if just the *X* or *Y* component of a vector were plotted as a function of angle. If the vector were rotating at a steady rate, the sine function would result if the *X* or *Y* component were plotted as a function of time. The concept of alternating current as the components of a rotating vector is valuable in the studies of certain types of circuits and radio-wave propagation.

To ably discuss circuit phenomenon for alternating current, some characteristics of the sine function must be understood. First, the sine function repeats a basic waveform and, essentially, goes on repeating forever. The number of times the sine wave repeats each second is called the *frequency* of the waveform and is given in *hertz* (Hz). Originally, frequency was given as cycles per second, which clearly defines the term. However, to honor Heinrich Hertz, an important physicist, the proper unit for frequency is hertz. The time between cycles is called the *period* and is the reciprocal of the frequency.

Example

What is the period of a 60-Hz wave?

Solution.   The period is the reciprocal of the frequency or 1/60 = 0.01667 or, more conveniently, 16.67 milliseconds (ms).

When discussing dc circuits, voltage is very simply defined. It is a constant and thus never changes. Such is not the case with ac. Since the voltage is constantly changing, we must be very careful when referring to ac voltage.

First, the voltage is a function of time and cannot be specified as a constant. A sine wave is described by the following time-dependent equation:

$$V(t) = A \sin(2\pi f)t \tag{2-10}$$

$V(t)$ is the voltage of an ac source as a function of time, t is time, and f is the frequency. The constant 2π is required because a sine function completes one cycle in 2π radians.

The peak amplitude of the sine function is A, which occurs when the sine function is equal to its peak value of 1. There is a negative peak, which is $-A$, that occurs when the sine function is equal to its negative peak of -1.

If we were to calculate the power delivered by an alternating current to a resistive load R, it would be obvious to assume that Ohm's law would provide the correct power. Therefore, an equation of the form

$$P = \frac{E^2}{R} \tag{2-11}$$

would be expected.

What is the correct value of E for calculating power using this formula? The voltage continually changes. We may be tempted to use A, the peak amplitude, but it is clear that this is not correct. The amplitude of the voltage source is equal to A only at the positive and negative peaks. At these instants, the power delivered to the resistance is A^2/R. At all other times the power delivery is less, sometimes as low as zero. There is a value of amplitude unique to an ac waveform that is between the peak amplitude and zero that will cause Eq. (2-11) to be correct. This ac amplitude is the *root mean square* (rms) value, and its definition is simple; it makes Ohm's law work. There are specific mathematical procedures for calculating the value of the rms amplitude for any waveform, but the end result is that the calculation of power using Ohm's law is correct when calculated with the rms value.

For a sine wave, or a cosine wave that has the same basic form as a sine function, the rms value is 0.707 times the peak value.

Example

What is the rms value of a 400-Hz sine wave with a 100-V peak value?

Solution.   The rms value for a sine wave is 0.707 times the peak amplitude, which is 70.7 V. Notice that frequency is not involved in the calculation of the rms voltage. It should also be understood that sine waves and cosine waves have identical shapes, but are offset by 90°. Therefore, the rms value of a cosine wave is exactly the same as for a sine wave.

The rms value is applied to both ac voltage and current. Essentially, all references to ac voltage and current are in terms of rms. As an example, if a piece of airborne equipment is to operate from 110 V ac, 400 Hz, the 110 V is the rms value and the peak voltage would be 155.5 V.

If rms values are used for both voltage and current, all the calculations involving Kirchhoff's and Ohm's laws with resistances are valid.

Why was the previous statement careful to state specifically "resistances"? Let us go back and have another look at the sine function, except this time we will look at two sine functions: one representing the voltage of a circuit and a second involving the current.

Figure 2-8(a) shows a case in which the voltage and current cross zero together and reach a peak value together. This represents the case where a circuit has only resistances and the voltage and current are said to be *in phase.* How is power delivered to the resistance of this circuit? The power is the product of the voltage and the current and, as can be seen from the figure, the power delivered to the resistance rises and falls with the voltage and current and rises and falls again in the second half of the cycle with the negative voltage and current. Thus, power is delivered to the load during both halves of the cycle, because both the voltage and current are positive in the first half-cycle and both the voltage and current are negative in the second half-cycle, again resulting in positive power.

Let us consider the case in which the voltage and current are not exactly in phase, as shown in Fig. 2-8(b). In this case the power does not always rise and fall with the voltage and current. Even more interesting, at certain points along the cycle the power is negative. If positive power represents an energy transfer from a source to a load, then negative power represents a transfer from the load to the source. For this to occur, the circuit must have some way of storing energy to be transferred back to the source. An electrical device capable of energy storage is called a *reactance.*

Electrical energy is stored in charges in one of two ways: moving changes and static charges. There is an analogy to the mechanical world, where energy may be stored in masses, with static mass containing potential energy and moving mass containing kinetic energy. Electrical charges that have energy content also give rise to fields. Static charges produce an electric field, whereas moving charges produce a magnetic field.

Generating an electric field requires that two conductors, ideally two conducting sheets, be separated by an insulator, called a *dielectric,* as shown in Fig. 2-9.

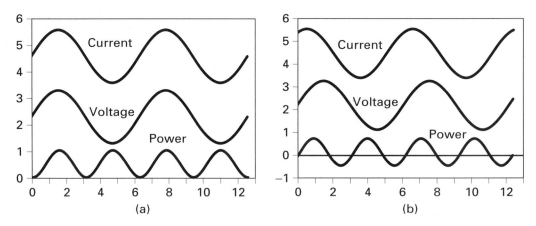

Figure 2-8   Voltage, current, and power relationships for two circuits. (a) Relationships for a circuit that contains only resistances. (b) Relationships for a circuit containing reactances showing the negative power flow.

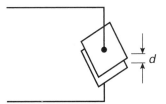

Figure 2-9 Basic construction of a capacitor.

If there is an electric field between the plates of a *capacitor*, to move a charge from one plate to the other against the electric field adds energy to that charge. Charges cannot jump from one capacitor plate to another because the material between the plates is an insulator. Charges can be transferred from one plate to another through an electrical circuit.

A parameter of two conductors separated by an insulator is called *capacitance* and is given by

$$C = \frac{Q}{V} \tag{2-12}$$

where C is the capacitance, Q is the charge that flows into the capacitor, and V is the voltage across the capacitor. This relationship is strictly an electrical relationship. There is no reference to physical dimensions or materials.

By a simple circuit analysis using Eq. (2-12), the energy stored in the capacitor as a function of voltage is

$$\text{energy} = \frac{1}{2} CV^2 \tag{2-13}$$

The capacitance of a parallel plate capacitor can be shown to be proportional to the area and inversely proportional to the distance between the plates and is also dependent on the material between the plates. These relationships should be intuitive. The larger the area is, the more charge that can be stored for the same charge density, and capacitance is proportional to the stored charge according to Eq. (2-12). The electric field between the plates of a parallel-plate capacitor can be shown to be a constant, so the voltage across the capacitor plates, V, can be shown to be

$$V = Ed \tag{2-14}$$

where E is the constant electric field and d is the distance between the plates. According to Eq. (2-12), capacitance is inversely proportional to the voltage and, thus, the capacitance is inversely proportional to the distance between the plates.

The capacitance of a parallel-plate capacitor is proportional to the area, inversely proportional to the distance between the plates, and a function of the material between the plates. Let us introduce a constant that provides the relationship between the material between the plates and the capacitance and write a relationship between capacitance, C, and the physical dimensions of a parallel-plate capacitor.

$$C = \frac{Ae}{d} \tag{2-15}$$

where A is the area of the capacitor plates in square meters, d is the distance separating the plates in meters, and e is *permittivity*, a parameter associated with the material between the plates. The value of e for air is 8.85×10^{-12} C^2/newton-meter^2.

Example

What is the capacitance of two plates with an area of 1 square meter (m^2) separated by 1 centimeter (cm) with air separating the plates?

Solution. The capacitance is found by inserting the values into Eq (2-15), which yields

$$C = \frac{(1 \text{ m}^2)(8.85 \times 10^{-12})}{1 \times 10^{-2}} = 8.85 \times 10^{-10} \text{ farads (F)}$$

A more common dimension for small capacitances is the picofarad (pF), or 1×10^{-12} farads. Therefore, the solution to this example is 885 pF.

Example

The previous example gives the impression that large capacitors are not practical. As an example, calculate the area that would be required to create a capacitor of 1 F using two parallel plates separated by 1 millimeter (mm).

Solution. Solving Eq. (2-16) for A, we have

$$A = \frac{Cd}{e} = \frac{1 \times 1 \times 10^{-3}}{8.85 \times 10^{-12}} = 1.13 \times 10^{8} \text{ m}^2$$

If this were created with a square metallic plate, the result would be a plate measuring 10,630 m on a side!

Capacitors of this size do not exist for two reasons. First, the separation of capacitor plates is considerably less than 1 mm. Second, capacitor values of 1 F are quite rare.

There are a number of ways to construct capacitors, and there is a wide variety of capacitors for numerous circuit applications. To achieve useful values of capacitance without involving large physical dimensions, many capacitors are made with very thin dielectrics that separate thin sheets of metallic foil. The sheets of foil and dielectric are wound in a cylindrical shape and covered with a protective outer coating. Capacitors of this style are called *plastic film capacitors* and are made of such dielectric foils as polyester, polypropylene, and polystyrene. Moderately large capacitor values can be obtained with this technique, and these capacitors are found in many lower-frequency circuits.

For higher frequencies, the winding of the foils causes undesirable effects. However, the foils are wound to save volume, particularly for the larger-value capacitors. Another method of reducing the size of a capacitor is to use a dielectric that has a high value of e.

By increasing the value of e in Eq. 2-15 and reducing d to the least practical value, physically small capacitors may be manufactured without winding. Ceramic materials have high values of e and are used to construct many high-frequency capacitors.

Capacitors are characterized by their capacitance, tolerance, and working voltage. *Working voltage* is that voltage that can be safely applied to the capacitor without breaking down the dielectric and damaging the capacitor. Dielectric breakdown occurs when a spark is generated and the heat from the spark melts or otherwise damages a part of the dielectric. Some capacitors can heal from a breakdown, but most capacitors are destroyed.

Another class of capacitors includes very high value capacitors. To obtain very high values, this class of capacitors uses an oxide layer grown on a metal foil as the dielectric. The oxide layer can be made very thin so that high-value capacitors may be constructed with reasonable dimensions. Like the plastic film capacitor, long lengths of foil may be wound to provide a compact capacitor.

These high-value capacitors have a unique characteristic. The oxide layer is grown on the metal foil through an electrochemical action. The oxide layer will remain unharmed as long as the electric potential across the oxide is of one polarity. If the capacitor is charged using the reverse voltage, the oxide layer may be harmed and the capacitor destroyed.

Capacitors of this sort are called *electrolytic* and use either aluminum or tantalum foils. These capacitors are made in very large values and are often used in power supplies.

Ceramic capacitors are typically manufactured in values from 1 pF or 1×10^{-12} farads to 0.01 microfarads (μF), or 0.01×10^{-6} farads. Ceramic capacitor working voltages can be from 50 to tens of thousands of volts. Plastic film capacitors are manufactured in values of about 1000 pF to 1 μF, with working voltages from about 50 V to about 500 V, with special variants operating at very high voltages up to 100,000 V. Tantalum electrolytic capacitors are made in values from about 0.5 to 330 μF, with working voltages from 6 to about 50 V. Aluminum electrolytic capacitors are made in values from about 1 μF to 1 F (or more!), with working voltages from about 5 to 1000 V.

Let us investigate the effects of capacitors in series and parallel. Consider the case of three capacitors connected in parallel as shown in Fig. 2-10. Since the capacitors are in parallel, the voltage across each capacitor is the same, and the total energy stored in the three capacitors can be obtained by applying Eq. (2-13).

$$\text{energy} = \frac{C_1 V^2}{2} + \frac{C_2 V^2}{2} + \frac{C_3 V^2}{2} \qquad (2\text{-}16)$$

If this same amount of energy is stored in only one capacitor with a value of C_{eq} with the same voltage, then we have the following relationship:

$$\frac{V^2}{2}(C_1 + C_2 + C_3) = \frac{V^2}{2} C_{eq} \qquad (2\text{-}17)$$

from which we can derive

$$C_{eq} = C_1 + C_2 + C_3 \qquad (2\text{-}18)$$

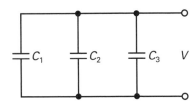

Figure 2-10   Circuit showing three capacitors in parallel.

Of course, this analysis can be extended to any number of capacitors, and the equivalent capacitance is simply the sum of the individual capacitors.

For capacitors in series, as shown in Fig. 2-11, we will write the equation for the total energy contained in the three series capacitors as

$$\text{energy} = \frac{V_1^2 C_1}{2} + \frac{V_2^2 C_2}{2} + \frac{V_3^2 C_3}{2} \tag{2-19}$$

Unfortunately, this relationship does not indicate how the voltage divides among the capacitors and we have no values for V_1, V_2, and V_3. To resolve this situation, let us recall our definition of capacitance:

$$V = \frac{Q}{C} \tag{2-20}$$

where V is the voltage across a capacitor, Q is the charge on the capacitor, and C is the capacitance. Using this relationship and realizing the fact that the charge transferred is the same for each capacitor since there is only one path for the charge to travel, Eq. (2-19) can be combined with Eq. (2-20) for the following result:

$$\text{energy} = \frac{Q^2}{2} * C_1 + \frac{Q^2}{2} * C_2 + \frac{Q^2}{2} * C_3 = \frac{Q^2}{2}\left(\frac{1}{C_1} + \frac{1}{C_2} + \frac{1}{C_3}\right) \tag{2-21}$$

From this relationship it can be seen that the equivalent capacitance of the three in series is

$$C_{eq} = \frac{1}{\dfrac{1}{C_1} + \dfrac{1}{C_2} + \dfrac{1}{C_3}} \tag{2-22}$$

where C_{eq} is the equivalent capacitance.

We showed how energy was stored in a capacitor by trapped charges on the capacitor plate. But, how can energy be stored in moving charges? Any length of wire that conducts an electrical current contains moving charges. These moving charges produce a magnetic field with magnetic flux lines oriented in concentric circles surrounding the wire (Fig. 2-12). Because each moving charge contains a specific amount of energy and each moving charge generates a specific amount of magnetic flux, the amount of energy stored by moving charges is related to the amount of magnetic flux produced. This is analogous to the electric field in a capacitor that contains a charge.

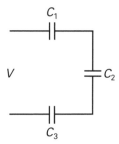

Figure 2-11   Circuit showing three capacitors in series.

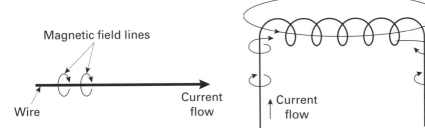

Figure 2-12 Magnetic lines of flux surrounding a
wire carrying a current.

Figure 2-13 Magnetic lines of flux of a solenoid.

If we wind the wire into a coil, or a *solenoid,* there would be a concentration of the magnetic field in the space surrounded by the turns of wire as shown in Fig. 2-13. This arrangement is the basic construction for an *inductor.*

As for capacitance in the case of a capacitor, a parameter called *inductance* and represented by the letter L will link the current through the inductor and the energy stored in the inductor as follows:

$$\text{energy} = \frac{LI^2}{2} \tag{2-23}$$

The unit of inductance is the henry (H).

The derivation of the actual inductance of a solenoid is beyond the scope of this text, but a very close approximation can be had from the following expression:

$$L = \frac{(N^2 d^2)\,10^{-6}}{18d + 40l} \tag{2-24}$$

where L is the inductance in henries, d is the inside diameter, l is the length of the inductor, and N is the number of turns in the solenoid.

Placing inductors in series or parallel is not a good practice, and therefore the derivations of series and parallel inductors will not be carried out. The reason for avoiding inductors in series or parallel is because of the uncontrolled interaction of the magnetic fields surrounding the inductors.

If two inductors are placed in close physical proximity, some magnetic field lines from one inductor may enter the other inductor and alter the magnetic field in that inductor. Just how much the magnetic field affects the inductance depends on the amount of magnetic flux leakage and the polarity.

Inductors take a variety of forms, with the simplest being a few turns of wire. Only a limited amount of inductance can be obtained with coils of wire. One problem with inductors constructed from wire solenoids is that they tend to allow magnetic field lines to stray far from the inductor. This causes energy to be lost to the surroundings of the circuit, which not only causes loss of energy, but can cause harmful interference to other circuits.

To create larger-value inductors without encountering physically large structures, ferromagnetic materials are introduced into the center of the solenoids. A ferromagnetic material has the characteristic that it tends to increase and concentrate the magnetic lines of flux. Therefore, when an inductor is wound around a ferromagnetic material, more lines of flux can be created for the same amount of current, and thus a solenoid with a ferromagnetic core can produce a greater amount of inductance for the same physical dimensions.

A parameter of the ferromagnetic material is called μ, which is the ratio of the flux lines in the material to the number of flux lines in a vacuum for the same coil current. This parameter is also the ratio of the inductance of the ferromagnetic material in the center of a solenoid to the inductance with air or a vacuum in the center. Therefore, for a solenoid wound on ferromagnetic material, Eq. (2-24) may be rewritten as

$$L = \frac{\mu \, (N^2 d^2) \, 10^{-6}}{18d + 40l} \tag{2-25}$$

where μ is the relative permeability of the magnetic material. The value of μ spans from about 50 or so to thousands. Most large inductors involve ferromagnetic material.

The concentration of the magnetic lines of flux helps to reduce the amount of magnetic field escaping from an inductor. But there is another very effective method of reducing the magnetic field radiation that involves the shape of the inductor.

Referring to the magnetic lines of flux from the solenoid inductor, the lines of flux leave the inductor at either end. Because flux lines have a closed path, lines leaving one end of a solenoid must travel around the outside of the solenoid to enter the other end. If the solenoid is bent around on itself so that one end is opposite the other in the shape of a donut, the magnetic lines of flux will be totally contained within the area encompassed by the wire, as shown in Fig. 2-14. This shape is called a *torus* and inductors wound in this fashion are called *torroidal inductors*. One significant disadvantage of a toroidal inductor is the cost of manufacturing because toroid winding equipment requires considerable manual operations.

Let us consider the current through an inductor. Inductors are made of wire wound into a solenoid or a torus with air or some type of core material. If we had taken just the wire without winding it into an inductor and connected it to a battery, the current would be limited only by the resistance of the wire. What would happen if the inductor were connected to the battery? Intuitively, the answer is that the current should be the same. Just winding a piece of wire in a solenoid or torus does not change the resistance of the wire, and this is absolutely correct. However, the winding of the wire into an inductor allows a much more concentrated magnetic field to be generated around the wire. The application of the wire to the battery causes the buildup of a magnetic field. But the building magnetic

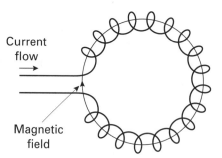

Current
flow
→

Magnetic
field

Figure 2-14 Toroidal inductors contain the magnetic field within the area surrounded by the turns.

field causes a current to be induced in the very wire that carries the current from the battery. This induced current is in the opposite direction of the current from the battery and reduces the current through the inductor.

The magnetic field cannot build forever, and eventually the magnetic field becomes a constant value, which results in no induced current in the wire. The current at this point is limited only by the resistance of the wire. Our intuition is right; the current, eventually, is limited by the resistance of the wire. The amount of time before the steady-state current is reached depends on the amount of inductance.

A similar situation occurs with the capacitor, but this involves the current. Since there is no current path through the capacitor, there can be no steady-state current. If a capacitor were suddenly connected to a battery, current would flow to charge the capacitor. Charges would move toward the negative plate and away from the positive plate. Don't forget that the moving charges are electrons and have a negative charge. The movement of charges is current, and the amount of charge movement, initially, is limited only by the resistance in the circuit. Eventually, the necessary charge has accumulated on the capacitor plates and the current ceases.

Application of a capacitor to a dc source causes an initial high current and eventually zero current, whereas the application of an inductor to a dc source causes an initial low current that eventually becomes a steady-state high current.

Although the leakage of the magnetic field from an inductor and the coupling of this field into other inductors are undesirable characteristics of inductors, this magnetic field coupling is used to advantage in a device called the transformer.

A *transformer* deliberately couples as much as possible of the magnetic fields of two coils as shown in Fig. 2-15. Consider that the first coil, the primary coil, is provided a changing current that is a sine function. The sine function is an excellent "changing" function because it is never stable. This function changes from zero to a positive peak and then immediately changes direction back through zero and to a negative peak and then back to zero. Then the process repeats.

The voltage induced in the secondary coil is a function of the primary. A very important characteristic of the coupled coils is used to create a very useful electrical circuit element. The voltage induced in the secondary coil per turn is the same for each turn of wire in the primary, as long as each turn encompasses the same magnetic flux. Therefore, the voltage across the entire secondary can be adjusted by adjusting the number of turns on the secondary. The coupled coils can "transform" a voltage level to a different value where the amplitude is proportional to the ratio of the secondary-to-primary turns, or

$$V_{\text{out}} = V_{\text{in}} \frac{N_2}{N_1} \tag{2-26}$$

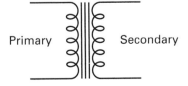

Figure 2-15 Schematic representation of a transformer.

where V_{out} is the secondary voltage, V_{in} is the primary voltage, and N_1 and N_2 are the number of turns on the primary and secondary, respectively. This ability to transform voltages has given the device its name, the transformer.

A theoretical transformer will transform an input voltage to an output voltage without any loss of energy. Therefore, if the input power is $V_{in} * I_{in}$, this is equal to the output power of $V_{out} * I_{out}$. For this relationship to hold, the current transformation is determined by the reciprocal of the turns ratio or

$$I_{out} = I_{in}\, \frac{N_1}{N_2} \tag{2-27}$$

where I_{out} is the output current and I_{in} is the input current.

Real transformers are not lossless, but the efficiency of a transformer is very good, being 99% or greater. Therefore, the relationships of Eqs (2-26) and (2-27) accurately predict the performance of real transformers.

The transformer operates only with alternating current. If there is no change of the voltage applied to the primary, there is no induced secondary voltage. If dc were applied to a transformer, as for an inductor only the resistance of the wire would limit the current, which would be very high.

To ensure that the maximum magnetic flux is coupled from the primary to the secondary, transformers use the flux concentrating characteristic of ferromagnetic materials. The primary and secondary coils are wound on the same ferromagnetic form called a *core*.

Transformers find a variety of uses in electronic circuits, particularly in power supplies. Universal disadvantages of transformers are their size and weight. Because the core is constructed of ferromagnetic material, which is very often steel, transformers tend to be heavy. By using a number of electronic methods, the use of transformers can be minimized, and those that must be used may be reduced in weight by employing special transformer steels. Transformers find widespread use in power supplies.

Example

What is the output voltage from a transformer operating at 400 Hz with an input voltage of 110 V ac, 400 Hz? The turns ratio is four to one from primary to secondary. What is the output power if the input current is 1 A?

Solution. The input 110 V ac is reduced by a factor of 4 to 27.5 V because of the four-to-one turns ratio. The output current is increased by a factor of 4. The input power is 110 W, which is identical to the output power. The 400-Hz input frequency is not involved in the calculation.

If a resistor were connected to the secondary of a transformer and the transformer connected to an ac voltage source, all the energy fed into the transformer will be delivered to the resistor. We assume for this example that the transformer is perfect and consumes no energy of its own. Therefore, the transformer connected to a resistor must be equivalent to a resistor alone because no energy is dissipated in the transformer. What is the value of the equivalent resistor?

Let us assume the voltage source is V_1, as shown in Fig. 2-16. The resistor connected to the transformer is R_2, and the transformer turns ratio is N. Therefore, the power delivered to R_2 is

$$P = \frac{(NV_1)^2}{R_2} \tag{2-28}$$

R_2 is the equivalent resistor if connected to the voltage source and would consume the same power, or

$$P = \frac{(V_1)^2}{R_1} = \frac{(NV_1)^2}{R_2} \tag{2-29}$$

Solving for R_1, we obtain

$$R_1 = \frac{R_2}{N^2} \tag{2-30}$$

This equation shows one application of a transformer, transforming resistance, which is properly called *impedance* for ac circuits.

Example

A pilot desires to use a pair of stereo headphones in an aircraft. The headphones have an impedance of 8 Ω. The headphones will be placed in parallel for monophonic use, and the pilot wishes to use a transformer to convert the impedance of the headphones to the 500-Ω output of a communications radio. What turns ratio is required?

Solution.   Because the headphones will be placed in parallel, the equivalent impedance will be 4 Ω. Therefore, the turns ratio is

$$\sqrt{\frac{R_2}{R_1}} = \sqrt{\frac{500}{4}} = \sqrt{125} = 11.2$$

The 8-Ω headphones with a 11.2-to-1 turns ratio transformer will perform as if they were a 500-Ω headphone set. If the 8-Ω headphones were connected directly to the radio designed for 500 Ω, the audio level may be too loud or too soft or possibly distorted.

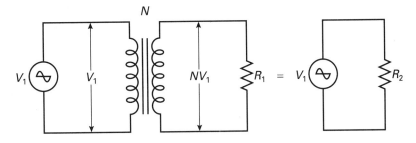

Figure 2-16   Transformer converting an impedance.

An important point must be made concerning the frequency of operation of a transformer. Many power supply transformers are designed to operate at only one frequency. Less steel is required of transformers operating at higher supply frequencies, and a transformer operating at a frequency other than the design frequency may not work properly. A transformer that is operating below its design frequency will likely suffer damage as well. Therefore, transformers that are designed, for example, to operate at 400 Hz may not be safely operated at 60 Hz. Since many power supplies use transformers, a power supply designed to operate at 400 Hz will suffer damage when operated at 60 Hz.

One of the more interesting phenomenon involving reactances is the concept of *resonance*. If a capacitor and an inductor are connected, energy can be transferred from one storage element to the other, for example, from the capacitor to the inductor. Consider the circuit of Fig. 2-17. Several observations may be made about this circuit. First, the circuit is capable of storing energy. Second, there cannot be a steady-state voltage across the inductor because, to dc, the inductor is a short circuit. Third, there can be no steady-state current through the capacitor because the capacitor is an open circuit to dc. Because the inductor and capacitor are connected, the entire circuit can have no steady-state current or voltage anywhere. Therefore, if any energy is contained in the circuit, the voltage and current must be in a continuous state of change.

By mathematical analysis, it can be shown that the current and voltage in the circuit are sinusoid in form and occur at only one frequency, the *resonant frequency*. The relationship of the resonant frequency to the circuit elements is given by

$$f_{res} = \frac{1}{2\pi \sqrt{LC}} \tag{2-31}$$

Because the circuit of Fig. 2-17 has only energy storage elements and no dissipative elements, once the circuit has an energy content, there will be no loss of energy and the circuit will oscillate indefinitely. Practical circuits are not lossless, and the energy content will diminish and the amplitude of the oscillations decay.

Resonant circuits that are less than perfect, where the energy eventually dissipates, are represented by the addition of a resistor to the resonant circuit.

A resistor is an example of an electronic circuit in which the transfer of energy is 100%. If a resistor R were applied to a voltage source of E volts, a current $I = E/R$ would result and power of EI watts would be transferred to the resistor.

The reactance is a situation in which a voltage E and a resulting current I can exist, but there is no permanent transfer of energy. The energy stored is in the reactance and there is no net transfer of energy. Therefore, power transfer is zero and not equal to EI.

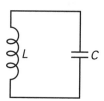

Figure 2-17   Resonant circuit using an inductor and a capacitor.

When circuits consist of a combination of resistance and reactance, there is some transfer of energy, but the amount lies somewhere between zero, the case of a pure reactance, and *EI,* the case for a resistor.

The relationship between the product of *EI* and the actual power delivered to a load is called the *power factor:*

$$\text{PF} = \frac{\text{power}}{\text{VA}} \tag{2-32}$$

where PF is the power factor and VA is the product of the supply voltage (volts) and the supply current (amperes). The power factor is the cosine of the phase angle between the voltage and current.

A unity power factor implies that the product of the voltage and current is equal to the power supplied to the circuit, which is a desirable situation. A power factor of less than 1 implies that, for the same power transfer, more current must be supplied. This requires a larger generating capacity only to increase the amount of heating of the connecting wires.

Example

A radar system operates from 400 Hz ac at 110 V with a power factor of 0.88 and requires 500 W. How much line current is required to operate the radar system?

Solution.   Equation (2-32) is rewritten to solve for the current, *A:*

$$A = \frac{\text{power}}{V * \text{PF}} = \frac{500}{110 * 0.88} = 5.17 \text{ A}$$

A total of 569 VA is required to operate the radar transmitter. If the power factor were corrected to unity, the VA required would exactly equal the power, 500.

The size of a generator is dictated by the VA requirement. A 569-VA generator would be required to operate the radar transmitter of the example, whereas a 500-VA generator would be required if the power factor were corrected.

Generators are big, expensive, and heavy. Therefore, every effort possible should be taken to achieve unity power factor for avionics equipment.

Active Circuits

With the exception of batteries, the circuit elements discussed so far, the resistors, inductors, and capacitors, are *passive* elements. This means that circuits constructed of passive elements can modify signals as they "pass" through the circuit, but they cannot generate signals or add energy to signals.

Devices that can generate signals or add energy to signals are called *active* devices. Active devices are mainly semiconductor devices and include such items as transistors, integrated circuits, silicon-controlled rectifiers, and field-effect transistors. There are literally thousands of different types of active devices. It is not possible in a text of this sort to even scratch the surface of the list of active device types, so only some of the simpler devices will be discussed.

Although they can add energy to a circuit, batteries are not included as active devices. The battery is a source of energy for a system and does not figure into any of the circuits other than as a source of power.

Before we can understand semiconductor devices, we must have an understanding of what a semiconducting material is. Previously, we discussed insulators and conductors and saw that insulators provide no passage for current, while conductors provide good passage for current. Resistors are somewhere between conductors and insulators. But how do semiconductors differ from resistors?

A *semiconductor* is more than a resistor because it provides current conduction in a very specific way. A semiconductor is a crystalline substance in which the specifics of conduction can be carefully controlled.

In a crystalline material, the atoms comprising the material are arranged in neat rows, rather than in a random fashion as in amorphous materials such as metals.

From our previous discussion of atoms, the atom was described as having a positively charged nucleus with orbiting electrons. The electrons orbit at only specific distances from the nucleus and, therefore, the electrons lie on a spherical surface called a *shell.* The atoms of higher atomic numbers have several possible shells, and the outer shell contains the *valence electrons.* These electrons are available for interaction with other atoms and for conduction of electricity.

One possible interaction of an atom with other atoms is the sharing of electrons in the outer shells of nearby atoms; this is called *covalent bonding.* Covalent bonding causes some atoms to align in regular shapes, such as cubes or diamonds, that are characteristic of a crystal.

Atoms have certain maximum numbers of electrons that may occupy a shell; the first shell can only have 2 electrons, the second 8, and the third 18. Silicon, atomic number 14, as an example has a full first shell with 2, a full second shell at 8, but only 4 electrons in the third shell, which is 14 short of being full. These 4 electrons in the outer shell are shared with the neighboring atoms in a covalent bond, which is the mechanism that holds the crystal in its regular shape.

Realizing that the four valence electrons are well bound to their neighbors, it would appear that pure silicon would not make a good conductor, and this is true. Imagine that some of the silicon atoms are replaced with an atom of arsenic, which has 5 valence electrons in its outer shell. Since all the silicon atoms are sharing 4 valence electrons and the arsenic atom is sharing 5, there is an extra electron that is free to drift. This drifting electron may be used to conduct electrical current, which allows the altered silicon to conduct more current than pure silicon. In addition, the amount of conduction can be set by controlling the number of arsenic atoms introduced into the crystal. The addition of the arsenic atoms is called *doping,* and silicon doped with arsenic is called N-type silicon, where the N refers to the negative nature of the conduction by electrons.

A similar situation exists when some silicon atoms are replaced with aluminum atoms, which have only 3 valence electrons. In this situation, most of the atoms, being silicon, are sharing the 4 electrons in their outer shell, but the aluminum atom is sharing only 3, which produces a void called a *hole.*

The hole causes the doped silicon to be a better conductor than the pure silicon in a fashion similar to the excess electrons in N-type silicon. Electrons can occupy a hole, but

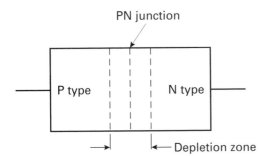

Figure 2-18   Semiconductor P–N junction.

this creates another hole, which effectively causes holes to move and current to flow. The holes, or lack of covalent bond, drift around just as do the excess electrons, but they have the equivalent of a positive charge. This type of semiconductor material is called P-type semiconductor, where the P refers to the positive nature of holes.

P- and N-type semiconductors have no interesting characteristics as just partially conducting material; it is when P- and N-type materials are brought together that very useful characteristics emerge. One semiconductor device is created when P- and N-type materials are brought together in what is called a *junction,* as shown in Fig. 2-18. At the boundary between the two different types of semiconductors, the holes of the P-type material are filled by the drifting electrons of the N-type material. Every time an electron fills a hole in the P-type material, it leaves the N-type material, which then has a more net positive charge due to the loss of the electron. As more electrons fill the holes in the P-type material, the more positive the N-type material becomes. This positive charge tends to attract the electrons filling holes in the P-type material, so after a number of holes are filled the attraction is so great that no further electron movement is possible to fill holes.

This results in a small, but finite area where there are filled holes and no free electrons. Empty holes and free electrons are the mechanisms whereby conduction took place in the semiconductor, but are now lacking in this area. Because this area has been depleted of its conduction mechanism, it is called the *depletion region.*

If the junction is connected to a voltage, it tends to force more electrons into the P-type region; these electrons can only fill more holes, further increasing the depletion region and preventing conduction. This voltage is called *reverse bias* and results in no current flow.

If a voltage of the opposite polarity is applied, it tends to push back the electrons that are filling holes and the depletion region decreases. If the depletion region reduces to zero, all the conduction mechanisms have been restored, and the junction can conduct. Conduction cannot occur until the depletion region has been reduced to zero, which requires about 0.7 V for silicon. Figure 2-19 shows the forward- and reverse-biased currents for a PN junction. Notice how a forward voltage of 0.7 V is required to cause conduction in the PN junction, and no current flows in the reverse-biased region. This device is called a *diode* and has the useful characteristic that current flows in only one direction. The schematic symbol for the diode is shown in Fig. 2-20(a). It is easy to remember that current can flow only in the direction of the arrow.

The diode has many applications in electronics, with one of the more important being as a rectifier for converting alternating current to direct current. Many electronic systems

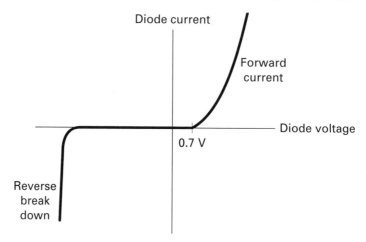

Figure 2-19   Forward- and reverse-biased characteristic of a PN junction device.

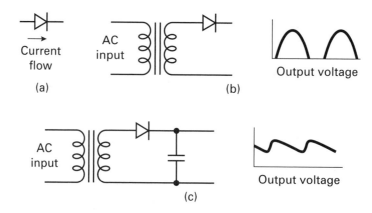

Figure 2-20   Transformer and diode power supply.

in larger aircraft are powered from a 110-V, 400-Hz ac source. Most electronic circuits require direct current for a power supply at voltages considerably less than 110 V. To provide this lower-voltage power supply, the 110 V ac is converted to a lower voltage using a transformer, and the alternating current is changed to direct current by passing the current through a diode, which allows current to flow in only one direction, as shown in Fig. 2-20(b).

The resulting voltage in Fig. 2-20(b) actually has only one polarity, but it is far from a steady constant voltage. We can now use the energy storage characteristics of a capacitor as shown in Fig. 2-20(c) to store energy and to provide a constant voltage between the cycles of ac.

Only one-half of the ac waveform is used in this example, and thus energy is transferred during one-half of the cycle. A much more efficient method would be to pass current during both halves of the cycle and pass twice as much energy each cycle. One common method for achieving this is to use a transformer with a split secondary, with one winding providing a sine wave exactly out of phase with the other. This type of transformer is called a *center-tapped secondary* and is shown in Fig. 2-21.

Using two diodes, current from the top half of the center-tapped secondary will pass through the top diode during the first half of the cycle. During this first half of the cycle, the diode connected to the bottom half of the transformer is presented with a reverse bias and does not conduct. During the second half-cycle, the entire situation is reversed; the top diode is reverse biased and the bottom diode is conducting. Therefore, energy is transferred during both halves of the cycle in the form of positive current into the load, and this type of arrangement is called a *full-wave rectifier*. The voltage from this rectifier still pulsates, but the amount of time that the capacitor must supply energy is considerably less and results in a more constant output voltage.

As simple as the diode is, a number of different types of diodes are used in electronic circuits. It is beyond the scope of this text to discuss every type of diode. However, one very important diode type needs discussion, the light-emitting diode or LED. This is one electronic component that almost everyone gets to see. The LED produces light when a current is passed through the diode. The current is in the forward direction because the LED is first and foremost a diode and passes current in one direction only.

The light from an LED is only one color, which is different than the light from an incandescent bulb, which is "white" light and contains a broad range of wavelengths or colors. LEDs are available in red, amber, yellow, green, and blue. The blue LEDs are the more difficult to produce and represent the state of the LED art.

Beyond the diode, the next most important semiconductor device is the *transistor*. There are many kinds of transistors made from a broad spectrum of materials. However, the most common transistor is the silicon bipolar transistor and it will be described here.

One common transistor is the *bipolar junction transistor*. The bipolar designation refers to the fact that the transistor contains both P- and N-type materials, usually silicon. The junction refers to the fact that the transistor has two junctions. The transistor provides current gain by changing the width of a depletion region, thus modulating the ability of the transistor to conduct and provide current gain.

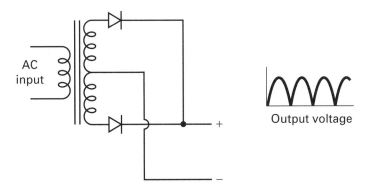

Figure 2-21   Transformer using a center-tapped secondary for a power supply.

The bipolar transistor is a current amplifier, and a simplified example of the bipolar transistor is shown in Fig. 2-22. There are three connections to the transistor, the base, emitter, and collector. If the transistor is considered as a simple amplifier, the emitter terminal is common to the input and the output, and the base is the amplifier's input, while the collector is the amplifier's output. This configuration is called a *common-emitter amplifier.*

The transistor can provide current gain or *beta,* depending on the type of transistor, from 10 to hundreds. Beta rightfully should be a negative number because current into the base causes an increased current into the collector. This means that the input device to the transistor provides a source of current while the output of the transistor amplifier is a current sink, and thus the common-emitter amplifier provides a signal inversion. The transistor amplifier is used in conjunction with other circuit elements to perform a large variety of functions and is the backbone of electronic systems.

Most electronic systems designed recently are based on integrated circuits. An *integrated circuit* places together on one piece of silicon, called a *chip,* transistors, diodes, resistors, and, to a smaller extent, capacitors. Tens of thousands of components may be placed on one silicon chip, which can represent a very complex circuit.

Integrated circuits are constructed on pieces of silicon called a *substrate* by selectively infusing the dopants and selectively creating P- and N-type silicon to create transistors, diodes, and resistors. Large numbers of transistors, in some cases over 100,000, may be manufactured on the silicon substrate. Silicon chips can provide the circuits for entire computers, radios, power supply regulators, and so on.

Electronic circuits can be categorized into two broad classifications, digital or analog. Analog signals can have any value, whereas digital signals can exist in only discrete values.

One example of an analog quantity is temperature. Temperature can assume any value and is not constrained to make $1°$ steps. An example of a digital quantity is the time as read from a digital clock. The clock provides the time in hours, minutes, and seconds. Time can be related with more precision than seconds, such as milliseconds or microseconds, but the clock doesn't provide this precision nor is it of any value for navigation purposes.

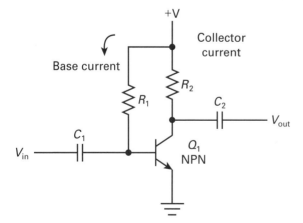

Figure 2-22 Bipolar transistor as a common-emitter amplifier.

Just as there are two broad classifications of electronic circuits, there are two classifications of integrated circuit, analog and digital. Analog integrated circuits include operational amplifiers, analog switching circuits, filters, multipliers, and a host of other applications-specific circuits. In the digital circuits, there is a very large selection of gates, counters, registers, flip-flops, and the like.

Analog quantities can be converted to digital by conversion techniques. And there are reverse conversion techniques that allow digital signals to be converted back to analog. These conversion techniques are called analog-to-digital and digital-to-analog or A–D and D–A conversion.

THE MICROPROCESSOR

The digital computer, particularly the microcomputer, has become an extremely important player in all phases of electronics. Because of this, A–D and D–A conversion has become an equally important technique. The A–D converter changes an analog quantity, such as the previous example of temperature, to a digital number that is processed by the microcomputer. The computer's result, which is digital, may be converted to an analog form if the end result requires it.

The advent of the microprocessor created a revolution in the electronics industry and certainly the avionics industry. To appreciate how the microprocessor had such an effect on the electronics industry, it is necessary to understand just what a microprocessor is. Quite simply, the microprocessor is a computer. It is not a computer that occupies an entire room, as many did years ago, not even a cabinet of equipment, but a single silicon chip or a handful of chips.

The basic requirements of a microcomputer are a central processing unit or CPU, read-only memory or ROM, random-access memory or RAM, and input/output devices or I/O devices.

The CPU is the brain of the computer. It is in this element that the necessary decisions are made to carry out the algorithms of the computer. An algorithm is the plan of attack for solving a problem using the computer.

The ROM contains the program and other fixed data the computer might need. If a computer is required to solve problems of geometry for navigation, the program for doing so is stored in the ROM and retrieved as needed. Other data may be stored in the ROM, such as the value of π or the number of degrees in a radian or other constants or even complete tables of data.

The RAM is used for storing temporary data that may be generated during the course of calculations. It may also be used for temporarily storing data that may be obtained more rapidly than they can be processed. If data in the RAM are no longer needed, the RAM can be reused for other data by simply writing the new data into the RAM.

The parts of a microprocessor system are tied together using data buses as shown in the block diagram of Fig. 2-23. Data flow in a microprocessor system is to or from the CPU and any of the peripheral elements. Generally, and there are exceptions to this rule, data pass from the addressed peripheral to the CPU where they are processed and transmitted to another peripheral.

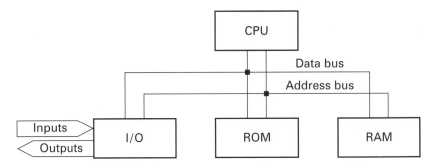

Figure 2-23   Block diagram of a microprocessor.

These data are transmitted on data buses. A bus is a set of wires that connects to the devices that will transmit or receive data. Each data word is a fixed number of bits and designates the number of wires in the data bus. Many microcomputers use an 8-bit data bus, which implies that data words are constructed in groups of 8 bits. Eight bits can only represent 256 different states and would not provide sufficient numerical range to provide the data needed for many calculations. When 8 bits does not provide the range necessary, the data word is divided into 8-bit groups called *bytes*. Large numbers are transferred on a data bus in these 8-bit groups.

The data bus of the microprocessor is a bidirectional bus, which means that data can flow from the CPU to a peripheral chip or from the peripheral chip to the CPU. Chips not transmitting data are isolated from the data bus by assuming a high-impedance state. This prevents the unselected chip from loading down the data bus.

In a microprocessor system, the bus is controlled by the CPU. Data placed on the bus by a peripheral chip go to the CPU, whereas data placed on the bus by the CPU go only to the chip addressed by the CPU. There is no peripheral-to-peripheral data transmission. The data buses used in a microprocessor are not transmission lines and data can only be transmitted for short distances.

The microprocessor also contains an address bus that is used to selectively enable peripheral chips as shown in Fig. 2-23. Each peripheral chip recognizes its address or block of addresses, and only the addressed chip responds with the requested data.

All the timing of the microprocessor system is through a number of synchronizing signals. One common signal is called the address latch enable or ALE. This provides a time reference for a valid address present on the address latch. Another important synchronizing signal is the read/write or R/W line, which synchronizes read and write operations. Depending on the microprocessor in use, different synchronizing signals may be present.

An important concept should be understood. A block diagram of a computer is the same for any other computer; they all have the same elements tied together with buses. What makes one computer solve mathematics equations while another analyzes a refrigerator is the *program*. What a computer does and how well it does it are functions of the program stored in the ROM. This is an important feature of the microprocessor, because improvements may be made to the computer by installing new ROMs rather than modifying circuits.

TRANSMISSION LINES

In our discussion of inductors and transformers, it was made clear that electric and magnetic fields are generated around wires carrying current. We used these fields to our advantage, but there could be situations where we would not appreciate the radiation of electromagnetic energy from wires carrying signals. To prevent this radiation (or the reverse situation, the ingress of energy into a circuit), transmission lines are used.

A *transmission line* is an arrangement of conductors by which the energy of the transmitted signal is contained within the transmission line. Energy is radiated when electric or magnetic fields are free to escape. One example of a transmission line is the coaxial cable shown in Fig. 2-24. This transmission line consists of a center conductor fully enclosed by an outer conductor. The electric field is fully contained between the two conductors, as shown. The magnetic field is fully contained by virtue of the fact that the current in the outer conductor is of exactly the same magnitude but in the opposite direction, and thus, to the outside of the transmission line, there is zero net current flow.

A second type of transmission line, the open-wire line, is shown in Fig. 2-25. This style of transmission line consists of two conductors that are kept parallel to each other throughout the entire length of the line. The transmission line is often twisted, which reduces signal egress or ingress and helps to mechanically hold the two conductors together. Transmission lines that are twisted in this fashion are often referred to as *twisted pair*. In this example, the electric field is not fully contained in the area between the conductors, but the field is very weak a small distance from the transmission line due to the fact that the current in one conductor is of exactly the same magnitude but opposite direction to the other conductor, so the net current at any distance from the transmission line is zero. For the open-wire line to function as a transmission line, neither side of the line may be connected to ground, and this type of connection is called a *balanced line.*

The twisted-pair transmission line tends to leak more energy than the coaxial cable, particularly at higher frequencies. One method of reducing this energy leakage is to enclose the twisted pair in a shield. The twisted pair is used for transmission of data when the frequencies involved are moderately low and some leakage of energy is acceptable.

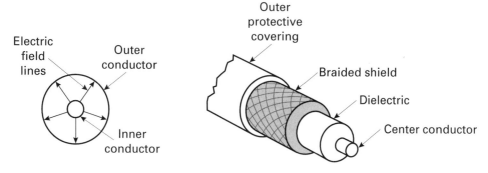

Figure 2-24   Coaxial transmission line.

Insulation

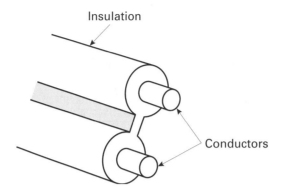

Conductors

Figure 2-25   Open-wire transmission line.

Coaxial cable is used for the transmission of radio frequency (RF) energy from a transmitter or to a receiver. The higher frequencies involved benefit from the superior shielding of the coaxial cable. In addition, most RF signals are not balanced but single ended.

Every transmission line has a very important attribute called the *characteristic impedance,* which is a function of the geometry of the line and the material of the dielectric.

Any electrical circuit will transfer a maximum amount of power only when the power source has the same resistance as the load. When power enters a transmission line, the maximum transfer of power occurs when the power source has the same resistance as the transmission line. The same occurs when power leaves the transmission line to a load. If the resistance of the load is not the same as that of the transmission line, the power transferred is not the maximum. When a mismatch occurs, there is literally power at the mismatched load that has nowhere to go. This power is reflected back through the transmission line to the power source.

This bidirectional transmission of power produces a number of undesired effects. First, if the transmission line is carrying data, data pulses will be traveling in two directions, and the reflected pulses will interfere with the forward pulses. In the case of continuous RF power, the reverse waves will add in or out of phase with the forward waves, depending on the length of the transmission line. This addition and subtraction of the forward and reverse waves give rise to standing waves. Standing waves inhibit the ability of the transmission line to transfer power. Therefore, to reduce the amount of data distortion and to maximize the amount of power transfer in a transmission line system, it is necessary to drive and terminate a transmission line in its characteristic impedance.

Most coaxial cables used in aircraft installations have a characteristic impedance of 50 Ω, and most twisted-pair transmission lines used for data transmission in aircraft have a characteristic impedance of 130 Ω.

To decrease the amount of energy radiated by a coaxial cable, some cables use two shields to increase the shielding effectiveness.

Energy can be lost from a transmission line not only from radiation but also due to the heating of the dielectric, particularly at higher frequencies. Therefore, most coaxial cables used at frequencies of 1 gigahertz (GHz) or higher not only use double shields but are made from low-loss dielectric materials such as tetrafluoroethylene, better known by its brand name of Teflon.

Antennas

We know from our investigations of inductors that wires carrying alternating current cause alternating magnetic fields and consequently a radiating electromagnetic field. The current-carrying element that converts conventional electrical current to a radiating electromagnetic field is called an *antenna*. Any current-carrying conductor will radiate, but some shapes and lengths serve the purpose better.

There are two basic types of antennas, electric field and magnetic field. This distinction is made relative to the generation of the electromagnetic wave. An electromagnetic wave is generated by creating either an alternating magnetic or electric field. A perpendicular field is automatically generated to make the propagating electromagnetic field.

Because of poor efficiency, magnetic field antennas are never used for transmitting and are only used for direction-finding receivers. This section will deal only with electric field antennas.

If a length of wire were fed with a current as shown in Fig. 2-26, the rising and falling voltage of the conductors would cause an electric field to exist between the conductors. However, the conductors do not represent a continuous circuit. Where does the current flow to return to the generator? Radio transmission involves propagating energy from a transmitting antenna, and this can only take place if power can be delivered to the antenna. If the length of each of the two conductors is a quarter-wavelength of the desired transmitting frequency, the antenna equivalent circuit is a pure resistance, which has nothing to do with the resistance of the wire of the antenna. This resistance is called the *radiation resistance* and represents the energy sink for the energy radiated into space. Remember that resistors, such as those used in a circuit, convert electrical energy to heat energy. The radiation resistance represents the resistance that converts electrical energy to radiating energy. This form of antenna is called a *dipole* and has a resistance of 50 to 60 Ω.

The dipole antenna will appear as a resistance at one frequency and odd harmonics of that frequency, which is called a *resonant antenna*. Like all resonant circuits, above and below the resonant frequency the antenna is reactive, meaning that the equivalent circuit is a capacitor or inductor with a resistance, the radiation resistance, in series. In the case of an antenna, the radiation resistance is also a function of antenna length. The shorter the antenna is, the smaller the radiation resistance becomes. Maximum energy is coupled to this

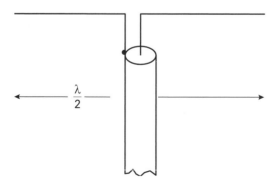

Figure 2-26   Dipole antenna.

antenna by providing an opposite reactance to counteract the reactance of the antenna and to provide a generator source resistance equal to the radiation resistance. An antenna of this sort is a nonresonant antenna and is still capable of radiating power. Nonresonant antennas are used for high-frequency communications when a resonant antenna would be too long to be practical. However, an antenna tuner is required to provide the series reactance and variable source resistance for the generator.

Example

What is the length of a resonant dipole for a frequency of 118 MHz?

Solution. The length of each half of the dipole is a quarter-wavelength, which is $(3 \times 10^8/118 \times 10^6)/4 = 0.636$ meter (m).

A modification of the simple dipole allows only one-half of the two radiators to be used. Only one-half of the dipole, called a *monopole,* is mounted perpendicular to the plane of a perfectly conducting surface called a *ground plane.* The conducting surface will provide the necessary current return for the current fed to the monopole. The result is a mirror image provided by the conducting surface. The mirror image is analogous to the missing half of the dipole antenna. Consequently, the antenna performs as a resonant antenna exactly like a dipole. Antennas of this sort are used on vehicles, especially aircraft, where the conducting ground plane is the vehicle body.

Spectrum

One important concept closely associated with radio transmission is that of spectrum. Frequency is involved in many phenomena, for example, resonance, wavelength, and modulation. *Spectrum* is the distribution of energy as a function of frequency. As an example of spectrum, assume that a radio receiver were used to record the signal strengths of radio stations in the standard broadcast band from 550 to 1700 kilohertz (kHz). The signal strengths would be plotted as a function of frequency in what becomes the spectrum of received signals for that frequency range.

The first characteristic of this plot would be that signals only exist at frequencies ending in even 10 kHz, representing a broadcast channel. (This would be true in the United States and most other countries. In some countries, particularly in Europe, the operating frequencies are 9 kHz apart.) Some signals would be strong, such as those from local stations or from high-power broadcast stations. Other signals from more distant stations would have a lower signal strength. Some 10-kHz channels would produce no strong signals, but a mixture of weak, distant stations that mutually interfere to produce unintelligible noise.

The signal from a broadcast transmitter is not a single frequency but occupies a band of frequencies. This is because the broadcast signal contains information that requires a band of frequencies to transmit. The higher the information rate is, the more frequency that is required to pass this information. In the case of the broadcast station, the information is speech and music, which have a bandwidth of about 5 kHz and require, at minimum, 10 kHz of radio spectrum. Therefore, each radio station requires a band of frequencies and a plot shows this rather than a point. Figure 2-27 shows an example of such a spectrum plot.

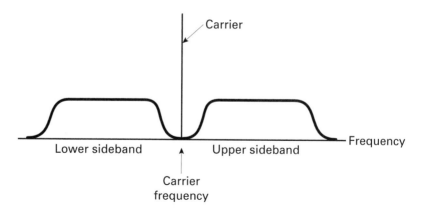

Figure 2-27 Spectrum of an amplitude modulated signal showing the upper and lower sidebands and the carrier.

TABLE 2-1

Frequency range	Nomenclature	Abbreviation
3 kHz–30 kHz	Very low frequency	VLF
30 kHz–300 kHz	Low frequency	LF
300 kHz–3 MHz	Medium frequency	MF
3 MHz–30 MHz	High frequency	HF
30 MHz–300 MHz	Very high frequency	VHF
300 MHz–3 GHz	Ultra high frequency	UHF
3 GHz–30 GHz	Extreme high frequency	EHF
30 GHz–300 GHz	Super high frequency	SHF

Some standard bands of the spectrum are handy to know. These bands span a decade, that is, a 10-to-1 ratio each, and are identified by wavelength. As an example, the band of frequencies corresponding to a wavelength of 1 to 10 m, or a frequency range from 30 to 300 MHz, is called very high frequencies or VHF. Many of these categories were named when the VHF range was thought to be very high. Relative to today's technology, VHF is not very high at all. The standard nomenclature is shown in Table 2–1.

MODULATION

The transmission of information by radio requires that the information to be transmitted be applied to the radio energy. The basic radio signal is called a *carrier* because it carries information. Modifying the carrier in some fashion to add the information is called *modulation*. A carrier is first and foremost a sine function, which may be given by the equation

$$S(t) = A \sin(2\pi ft + \theta) \tag{2-33}$$

There are only three parameters that may be modulated to apply the information; the amplitude A, the frequency f, and the phase angle θ. Time, t, cannot be modified because it is a universal invariant. All three forms of modulation can be and are used in aviation systems.

Amplitude modulation is used for VHF communications and involves the variation of the carrier's amplitude in direct relationship to the information. Let us investigate a simple form of modulation, a single sine function that could be a steady tone such as a whistle or the Morse code identifier for a VOR transmitter.

The amplitude of the carrier would rise and fall with the sine function as shown in Fig. 2-28. At the minimum amplitude of the modulating sine wave, which is actually a negative one, the carrier would be at a minimum or zero amplitude. This represents 100% modulation, the maximum achievable with amplitude modulation. Since the carrier amplitude cannot go below zero, any attempts at increasing the percentage of modulation beyond 100% will result in a flattened negative peak and in distortion of the modulation waveform.

The equation of the amplitude-modulated carrier is given by

$$[1 + MF(t)] \sin (2\pi ft) \tag{2-34}$$

where M is the modulation index or the percentage of modulation divided by 100, $F(t)$ is the modulating function with a maximum value of 1, and f is the carrier frequency.

Even though the carrier receives the modulation, whether there is amplitude modulation or not, the carrier is always there. When a carrier is amplitude modulated, the energy added to the carrier is distributed above and below the carrier frequency in two energy bands called *sidebands*. For amplitude modulation, both sidebands contain the same information. Therefore, for amplitude modulation there are three components, the carrier, which carries no information, and the upper and lower sidebands, which, although they carry the information, are redundant.

To reduce the amount of spectrum required by amplitude modulation and to reduce the power necessary to transmit information, a modulation technique called *single sideband* is used. In this modulation technique, the carrier and one of the two sidebands are removed, leaving only one, either the upper or lower, sideband.

Another method of modulation is achieved by changing the frequency of the carrier in step with the information; this is called *frequency modulation.* Since frequency is the change of angle in the carrier function, modulating the final parameter, phase, actually produces frequency modulation. There are some significant advantages of frequency modula-

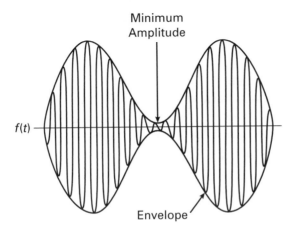

Minimum
Amplitude

$f(t)$

Envelope

Figure 2-28 Envelope of an amplitude-modulated carrier.

tion, but because this form of modulation is not used in aircraft systems it will not be covered in this text.

If there are advantages to using frequency modulation, why isn't it used for aircraft communications? The reason is more historic than practical. When aircraft were first equipped with communications capability, in the mid 1920s, frequency modulation had not been invented. In fact, even though theoreticians knew that frequency modulation was possible, they could not agree on the potential advantages of the technique.

DECIBELS

The decibel (dB) is a very important method of specifying voltages, currents, power, gains, and losses in electrical systems. A *decibel*, which is actually one-tenth of a bel, was named for Alexander Graham Bell. The bel is an inconvenient measure and only decibels are used. To define a ratio of power as a decibel equivalent, the following formula is used:

$$dB = 10 \log \frac{P_2}{P_1} \tag{2-35}$$

where P_1 and P_2 are two powers. What is the attraction of specifying the ratio of two powers as a decibel amount?

First, why are we concerned with ratio of two powers? As an example, the ratio of the output power to the input power of a circuit indicates the gain or loss of that circuit. If the power output of a circuit is 10 watts (W) while the input power is 1 W, the power gain of the circuit is 10, which also happens to be 10 dB. Let us assume that two of these circuits are placed in tandem. Therefore, only 0.1 W is required to provide a 10-W output. The total gain is 10×10 or 100. The gain in decibels is 10 dB + 10 dB or 20 dB.

If P_2 is the output power of a device and P_1 the input power, and if P_2 is greater than P_1, the device provides a gain and the decibel value for this device is a positive number. If the output power is less than the input power, the device has a loss and the decibel value for this device is a negative number.

When circuit elements are placed in tandem, to find the equivalent gain or loss, the decibel gain or loss of each element is summed.

Example

What would be the total system gain if the following elements were connected in tandem?

$$A = 12 \text{ dB}$$
$$B = -6 \text{ dB}$$
$$D = 10 \text{ dB}$$
$$E = 14 \text{ dB}$$

Solution.

The sum of the gains or losses is 30 dB.

To find the gain or loss as a ratio, Eq. (2-36) is rewritten as

$$\frac{dB}{10} = \log \frac{P_2}{P_1} \qquad (2\text{-}36)$$

Both sides of the equation are raised to the power of 10:

$$10^{dB/10} = 10^{\log(P_2/P_1)} = \frac{P_2}{P_1} \qquad (2\text{-}37)$$

Example

What is the equivalent power gain of a 12-dB circuit?

Solution.

$$\frac{P_2}{P_1} = 10^{1.2} = 15.8$$

Decibels are a convenient method of handling ratios, or gains and losses. If P_1 were a reference power such as 1 milliwatt (mW), decibel notation could be used to express an absolute power as a decibel figure relative to the reference power. A very common standard notation is the dBm, which is a power expressed in decibels relative to 1 mW.

Example

What power is a signal with a level of +23 dBm?

Solution. In this case, P_1 is 1 mW and P_2 is the power in question. Therefore, solve Eq. (2–38) for P_2.

$$P_2 = P_1(10^{2.3}) = 1 \times 10^{-3} \times 200 = 200 \text{ mW}$$

Powers expressed in positive dBm numbers are greater than 1 mW, while powers represented by negative dBm figures are less than 1 mW. Zero dBm is 1 mW.

The use of dBm makes the calculation of signal levels simple when used with gains and losses expressed as decibels. To find an output power level, add the gain or loss, remembering that the loss will be adding a negative number to the input power in dBm to compute the output power in dBm.

Example

What is the output power if a −15-dBm signal is applied to a tandem arrangement of the following gains and losses?

$$A = -3 \text{ dB}$$
$$B = 25 \text{ dB}$$
$$C = 14 \text{ dB}$$
$$D = -6 \text{ dB}$$

Solution. The gains and losses sum to 30 dB, which when added to −15 dBm is +15 dBm.

Many specifications for radio-type navigation systems are based on dBm, and its use is nearly universal for measuring radio-frequency power.

REVIEW QUESTIONS

2.1. The identification tag for a window heater for the pilot's windscreen rates the heater at 125 W operating at 28 V.
  (a) What current is required for this device?
  (b) What is the equivalent resistance when operating?

2.2. An ac operated electrical unit requires 100 VA from a 400-Hz, 115-V source with a power factor of 0.88. How much power does the unit consume?

2.3. The equipment in Problem 2.2 requires a circuit breaker with 1.5 times the current draw. What value circuit breaker must be used?

2.4. A 200-Ω resistor is connected to a 115-V ac, 400-Hz source. How much power is dissipated by the resistor? How much energy is provided by each cycle of the 400-Hz source?

2.5. What is the peak voltage of a 220-V, 60-Hz sine wave?

2.6. What is the capacitance of the series combination of the following capacitors: 0.1 μF, 0.33 μF, 0.47 μF, and 0.68 μF?

2.7. What type of capacitor is used when very high capacitance values are required such as for a power supply?

2.8. What characteristic of a resistor affects the size of a resistor?

2.9. What is the value of three resistors of 100, 220, and 330 Ω connected in parallel?

2.10. What is the value of resistance of the resistors in Problem 2.9 connected in series?

2.11. When connected to a 28-V dc source, which would dissipate the most power, three 100-Ω resistors in series or in parallel?

2.12. How much energy does a 12-V, 72 ampere-hour battery contain when charged to three-quarters capacity?

2.13. An emergency locator beacon has a 9-V, 5200-mA-hour lithium battery. The transmitter consumes 1.75 W and transmits continually until the battery runs down. How long will the transmitter operate?

2.14. What is the wavelength of a 75-MHz marker beacon signal?

2.15. What is spectrum?

2.16. If a 20-V carrier is modulated with 75% modulation, what is the peak carrier voltage?

2.17. If a signal has an amplitude of 112 μV and is passed through an amplifier that raises the amplitude to 500 μV, what is the gain of the amplifier in decibels?

2.18. A transponder transmitter operates at 125 W. How much power is that in dBm?

2.19. An older receiver sensitivity specification was for a signal level of 10 μV working with a resistance of 50 Ω. The new specification calls for the same power, but the level is given in dBm. What is the power in dBm?

3

Aircraft Electrical Systems

In this chapter the student will learn the fundamentals of power generation and distribution. The student will gain an understanding of the characteristics of interconnecting wires and connectors. This chapter also covers aircraft lighting systems and static inverters. Any entity that uses electrical power must have a safe and effective method of distributing that power.

INTRODUCTION

Most of us take for granted that electrical power will be available whenever we wish to turn on a light or other appliance in the home. This reliable availability of power is achieved through a very sophisticated system of power generation and distribution operated by the utility companies. Having a reliable source of power in an aircraft is just as important and it, too, relies on an effective method of power generation and distribution.

To start the discussion of power generation and distribution in an aircraft, the general characteristics of the major power sources will be discussed. Two basic types of power are available on board an aircraft, ac and dc.

Direct current (dc) is the one and only form of power in a light aircraft because storage batteries are an effective way of storing energy to be used before the engines are operating. To simplify things, only dc is used to power all the devices aboard the aircraft. Although various voltages were used in older aircraft installations, the more common were 14 and 28 volts (V). The higher 28 V has emerged as the only voltage for new aircraft.

The dc source of energy was ideal for motors, lights, actuators, and other devices, but was a decided disadvantage for early radio equipment, which required high voltage for its vacuum tubes. To generate the required high voltage, a variety of methods were used, many of which employed large mechanical components such as motor–generator sets.

In larger aircraft, an ac source of power was supplied. This eased the generation of the high voltages required for radio equipment because simple transformers and rectifiers were used. To save weight, 400-hertz (Hz) power was employed to reduce the size of the required transformers. The standard voltage for the 400-Hz power supply was set at 115 Volts, which was simply the same as used in common 60-Hz ground power sources.

For any ac power distribution system, a certain amount of energy is transferred for each cycle of the ac power supply. The actual instantaneous delivery of power rises and falls during each cycle. In the case of 60 Hz, this is often apparent as a vibration in some equipment, particularly motors. In the case of a 400-Hz supply, the vibration is a higher frequency but is still present. The reason for the unsteady delivery of energy is that the sine function that delivers the power rises to a peak and falls to zero twice every cycle. At the zero crossings, there is no delivery of energy. A solution to this problem is to provide three sine waves such that, when one sine wave is at zero, the other sine functions are not. The sine functions are of exactly the same frequency but with a constant phase angle difference. Any number of sine-wave voltages will provide the desired improvement, but power generation and distribution for both aircraft and ground use have standardized on three phases with a 120° phase relationship between the three.

For powering large equipment, three-phase power is available on large aircraft. Three-phase power provides three circuits each with a 400-Hz sine wave, but a 120° phase angle between the sine functions. Therefore, for each cycle of 400-Hz power, there are three positive peaks and six negative peaks. There are a number of advantages for this type of power distribution. One is that more efficient power supplies may be made for generating high voltage for vacuum-tube circuits. Another advantage is that large motors run more smoothly when they use multiphase power.

Another use of 400-Hz power is to drive indicators and displays. Precision motors, called *servo motors,* are used to sense angular position and to move indicators for displays. Servo motors are also used as position sensors associated with controlling aircraft surfaces, such as with an autopilot or a flap-setting system. Many of these motors are small and do not require the full 115 V supplied by the ac power distribution system. For these servo motors, a lower-voltage, 400-Hz source was made available with voltages of 13 or 26 V. If such a system is available, it is not used for providing power for electronic equipment but for driving the small servo motors.

Providing a constant-frequency ac power source requires that the generator turn at a constant rpm. Because the aircraft's propulsion engines operate over a range of rpms, when a 400-Hz generator is driven from one of the engines, a constant-speed drive unit is used to stabilize the speed of rotation of the generator.

Larger aircraft have an *auxiliary power unit* or APU turning at constant rpm to generate the 400-Hz power required by the aircraft before the aircraft is airborne. Typically, an APU is a small turbine turning a 400-Hz generator, either single or three phase, and often both. In addition, some APUs provide a dc generator to charge the aircraft battery and hydraulic pumps for energizing hydraulic systems.

In smaller aircraft, the only source of power is dc, and an engine-driven generator is used to charge the battery. The speed of the engine is not an important factor. Unlike the ac power source, for which the frequency of the power is critical, the dc generator requires regulation of the output voltage, which is also a function of the speed of the generator, but can function at a variety of shaft speeds. When the engine is running at idle, the equipment is powered from the battery, and when the engine is operating at normal cruise rpm, the energy in the battery is replaced.

Since vacuum-tube equipment is, essentially, gone from modern aircraft, generating high voltages for radio equipment is no longer a serious problem, and most modern radio equipment operates with internal voltages of 28 V and below. When higher voltages are required, modern power-conversion techniques can generate the needed voltage efficiently and with low weight. Therefore, light aircraft can be outfitted with very sophisticated avionics without the use of 400-Hz power, with the exception of a small amount of 400-Hz power to operate the instrument servo systems.

DC POWER SYSTEMS

Every aircraft has a dc power system, even the largest air transport aircraft in which practically everything operates from the 400-Hz power source. The dc power system is used to provide emergency power and to power those devices that are required when the APU is not running, such as the starter motor for the APU. In smaller aircraft, the dc power system is the only source of power. Whether large or small aircraft, the dc power is generated and distributed in the same way. There are four major components of the dc power system: the generator, battery, voltage regulator, and distribution system.

Batteries

Every aircraft contains at least one battery. In a small aircraft, the battery supplies the necessary power to start the engines and provides power for the avionics, lamps, and other items before the engines are started.

Many circuits shown in Chapter 2 included a voltage source, which was a perfect or ideal battery. The perfect battery has some worthy characteristics: it never goes dead; the voltage is a constant and never changes as the battery becomes discharged; the amount of energy available from the battery is infinite; and so on. Real batteries are less than perfect, and a good understanding of the real characteristics of aircraft batteries is necessary. The schematic diagram of a real battery is shown in Fig. 3-1.

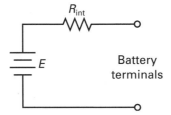

Figure 3-1   Equivalent circuit of an aircraft battery.

The first thing we notice about the real battery is that it has an internal resistance, R_{int}, which implies that infinite current is not available from the battery. A second fact that is not evident from the equivalent circuit is that the internal resistance and the maximum voltage available from the battery are functions of the state of charge, the temperature, and the age of the battery. Even though the battery symbol was used as a steady source of voltage for circuit analysis, the real battery is no such thing.

There are two basic categories of batteries: primary and secondary. Primary batteries are the types that cannot be recharged. Once the energy is depleted, the batteries must be discarded and replaced with new. Batteries of this type are the zinc–carbon, lithium, or alkaline types. These are the batteries used in flashlights or portable equipment.

A point of terminology should be mentioned here. The term *battery* refers to several *cells* connected together. The flashlight "battery" is actually a flashlight cell. However, once two or more cells are placed in the flashlight, the flashlight now contains a battery. It is improper to say that two batteries were inserted into the flashlight. However, few people make this distinction and it is becoming acceptable to refer to individual cells as batteries.

A secondary battery can be recharged, which allows the energy that has been removed to be replaced. The major application of a secondary battery in an aircraft is for the storage of energy as dc electrical potential energy. The source of this energy is an engine-driven generator.

Of the secondary battery types, only two are in use in aircraft. The lead–acid battery is similar to those commonly found in automobiles and is also the most common in aircraft. The nickel–cadmium battery is considerably more expensive and is found in some aircraft, particularly turbine-powered aircraft.

The lead–acid cell consists of a group of lead plates that are immersed in an electrolyte. The positive plates are made of lead peroxide and the negative plates are pure lead. The electrolyte is 30% sulfuric acid and 70% water. The lead peroxide and pure lead are very soft, and usually the lead is held in a grid made of much more rigid materials that will not affect the battery's chemical reactions and will provide a low-resistance current path for the generated current.

Because the electrical current provided by the battery must actually travel through the electrolyte, this distance must be kept to a minimum so that the battery will have the highest efficiency. With this distance at a minimum, only a small amount of warping of the battery plates will cause the plates to touch. To prevent the plates from warping and touching, thus causing a short, insulating material is inserted between the plates. This material must be a good insulator, must be immune to chemical reaction from the sulfuric acid electrolyte, and must not be involved in the chemical reaction of the battery. In addition to the warping of the plates, during the lifetime of a lead–acid cell some of the lead may flake off the plates, and if these lead flakes become dislodged between the plates, they can cause a short. The insulating material holds the debris away from the plates and prevents the development of these shorts. The insulating material placed between the plates is called a separator and is typically glass fibers or various types of plastics.

In spite of the separators, some small bits of debris will flake off the plates and filter to the bottom of the battery. To prevent this debris from shorting the plates, a space is provided at the bottom of the battery case so that the debris can accumulate away from the plates.

During the charging of a cell, particularly at the end of the charge, gas is emitted from the electrolyte. This causes two problems. First, the gas must be vented to prevent the buildup of pressure within the battery case. Second, the venting of the gas causes a loss of electrolyte because the gas is disassociated water from the electrolyte. Therefore, there must be a method of replenishing the electrolyte. Thus, the battery is fitted with access holes with threaded caps with vent holes that allow the gas to escape and provide an access for adding water.

In an aircraft battery, more so than in an automobile battery, care must be taken with the design of the battery caps so that gas may escape but the electrolyte does not escape. The vented gases and electrolyte are both dangerous. The gases are oxygen and hydrogen, which constitute an explosive mixture. The vented gases must be free to dissipate, and thus battery compartments must have adequate ventilation. The electrolyte is highly caustic and should not be allowed to escape.

Battery caps have a number of mechanisms for allowing the gases to escape but preventing the electrolyte from escaping. Typical techniques involve tilt valves that close off the vents if the battery is tilted or porous material that will allow the gases to escape but present a barrier to the larger electrolyte molecules.

The chemical action of the battery takes place on the surface of the plates, and thus the maximum surface area is desired. To maximize the surface area, the plates are as thin as practical, and the space between the plates is the minimum practical. The energy storage capacity of a lead–acid cell is proportional to the amount of lead with its surface exposed to the electrolyte. An important battery parameter is the energy-to-weight ratio. To maximize this ratio, the least amounts of lead, electrolyte, and other battery materials are used.

The electrical energy placed into the lead–acid cell during charging is stored as a chemical reaction. When current is permitted to flow, which is the transfer of electrons, the electrons can pass from one plate to another, but current can also pass through the electrolyte. The electrolyte disassociates into hydrogen ions, H_2, which have a positive charge, and sulfate ions, SO_4, which have a negative charge. The negatively charged SO_4 ions combine with the lead plate to form $PbSO_4$. This results in the SO_4 ions giving up their electrons, which frees those electrons for conduction.

The positively charged H_2 ions travel to the lead peroxide plate, combine with the oxygen atoms in the PbO_2 molecule to form water, and give up their positive charge by taking electrons from the lead peroxide.

Therefore, excess electrons appear at the lead plate while the electrons are removed from the lead peroxide. This lack of equilibrium is neutralized by providing a current path between the plates. This current path is through the load, which equalizes the charge of the plates and permits the chemical reaction to continue and provide electrical energy.

The chemical reaction at both plates produces lead sulfate, $PbSO_4$, which isolates the plates from the electrolyte. This prevents the chemical reaction from taking place, and the cell is discharged and no further current may flow.

When current is passed in the opposite direction, the lead sulfate is decomposed by driving the SO_4 ions back into the electrolyte. Therefore, the original lead and lead peroxide of the plates are restored, and the cell is capable of producing a current flow once again.

Another type of secondary cell used in aircraft power systems is the nickel–cadmium cell. This cell uses plates of cadmium for the negative plate and nickel oxyhydride for the positive plate. The electrolyte is 70% water and 30% potassium hydroxide. The construction of the nickel–cadmium cell is similar to the lead–acid cell using separators.

The nickel–cadmium battery does not vent as much gas as the lead–acid cell unless the cells are overcharged. Therefore, the cells may be sealed with a safety pressure valve that will vent only when the internal pressure has built up due to extended overcharging. Because this venting of gas is not from the disassociation of the water in the electrolyte, the nickel–cadmium cell requires no replenishment of the water unless it has been significantly overcharged.

Each lead–acid cell has an open circuit voltage of about 2.1 V, which drops down to about 2.0 V under a normal discharge. This voltage is a function of state of charge and temperature. Therefore, to generate a 24-V battery, 12 cells are required with a no-load voltage of 25.2 V. The nickel–cadmium cell has an open-circuit voltage of about 1.2 V, which changes very little between no load and full load. The nickel–cadmium battery requires 20 cells to make a 24-V battery.

One of the more important specifications for a battery is its energy capacity, which is usually given as an ampere-hour (A-h) rating. Essentially, this is the amount of current a battery will provide for 1 hour (h); it is also called the 1-h discharge rate.

The actual energy that may be stored in a battery depends on the rate at which the energy is removed. When a battery is discharged rapidly, because of internal heat buildup, the amount of energy available from the battery is reduced. On the other hand, batteries have some self-discharge due to small internal leakage currents. If a battery is discharged over a long period of time, the leakage current may discharge a significant amount of energy. Therefore, there is an optimum discharge rate between excessive heat buildup and the self-discharge rate at which a battery provides the maximum output energy.

Example

A 24-V aircraft battery has an ampere-hour rating of 50 A-h. What is the energy capacity of the battery in joules?

Solution.   Since the battery would supply 50 amperes (A) for 1 h, the power delivered by the 50 A would be 1200 watts (W) and would last for 3600 seconds (s). This represents a total energy of 4.32 million joules (J).

Example

How long would a 35-A-h battery power a communications radio during an emergency situation if the radio required 1.5 A? Assume the battery was fully charged before the radio was used.

Solution.   The total time in hours is $35/1.5 = 23.33$ h. This solution ignores the self-discharge current, which, for a battery in good condition, is usually considerably less than 1.5 A.

Aircraft batteries are seldom run down so that they are completely discharged. The battery is used for a short period of time before the engine is started and to power the starter motor. Once the engine is operating, the battery immediately begins recharging to a full charge. When a battery has been run down, it is not desirable to recharge the battery at an excessively rapid rate. This causes unnecessary stresses on the battery and may damage it. Typically, a battery is recharged at one-tenth of the 1-h discharge rate. Also, since the battery is not efficient, 60% additional energy must be placed into the battery. Therefore, the battery must be charged for 16 h at one-tenth of the 1-h discharge rate to be fully recharged.

Example

> At what maximum rate may a 50-Ampere-h battery be charged and for how long to fully charge a battery with half a charge?
>
> **Solution.**   The maximum charge rate is one-tenth the 1-h discharge rate or 5 A. The total charge that has to be replenished is 25 A-h, and, because of the inefficiency of the charging process, we use 1.6 times 25 A-h, or 40 A-h. At a 5-A rate, this requires 8 h.

Some characteristics of battery charging, are important. Refer to Fig. 3-1, the equivalent circuit of a battery. Because there is an internal resistance within the battery, the voltage at the battery terminals is less than the open-circuit voltage of the battery when current flows out of the battery. Also, when charging current is flowing into the battery, the terminal voltage is greater than the open-circuit voltage.

A 12-cell lead–acid battery providing a load at one-tenth the 1-h discharge rate would have a terminal voltage about 2 V below the nominal 25.2 V, or 23.2 V. When the same battery is being charged at the one-tenth rate, the terminal voltage is 2 V above the nominal 25.2 V, or 27.2 V. Therefore, in normal operation the battery terminal voltage will span from 23.2 to 27.2 V. If the battery were discharged or very cold, this variation would be even greater. It is clear that the battery schematic symbol that represents a constant source of voltage does not begin to represent a real battery as found in an aircraft. Because most operation of aircraft electronics equipment takes place while the engines are operating and the batteries are under charge, most avionics equipment intended to operate with 12 lead–acid cells, a "24-volt" battery, is designed to operate with a nominal 27.5- or 28-V power source.

The output voltage of both the lead–acid and nickel–cadmium batteries decreases with reduced charge. The nickel–cadmium battery tends to maintain a constant voltage and suddenly fall near full discharge. The lead–acid battery tends to fall off gradually.

The internal resistance of the batteries tends to increase as the battery is discharged. The nickel–cadmium battery has a lower internal resistance that changes little as a function of state of charge. The lead–acid battery tends to have an internal resistance that steadily increases during the discharge process.

The lower and more constant internal resistance and the more steady output voltage during discharge are two of the advantages of using a nickel–cadmium battery in an aircraft. Also, the reduced venting represents a lower-maintenance battery, and the battery has a longer life. The energy-to-weight ratio of the nickel–cadmium battery is higher, which represents a weight saving. With all these advantages, why is the nickel–cadmium battery used only for starting turbine-powered aircraft? There is one major disadvantage of the nickel–cadmium battery and that is the cost. The nickel–cadmium battery can be five times more expensive than a comparable lead–acid battery.

Power Generation

The generator is the source of electrical energy for the dc power system. The generator must be provided with a source of mechanical energy, which is from the propulsion engines or an APU.

The connection between mechanical energy and electrical energy is the current induced in a conductor that is passed through a magnetic field. The source of mechanical energy available is the rotation of a shaft from an engine. Therefore, a rotating coil or wire is used as the basis for electrical power generation.

Figure 3-2 shows the simplest form of electrical generator, which is a rotating coil in a fixed magnetic field. The magnetic lines of flux pass first through the coil in one direction and then on the other half of the rotation cycle pass through in the opposite direction. Because the magnetic field first passes in one direction and then the opposite, the polarity of the induced current alternates. Thus, the output current is alternating or ac.

A generator inherently creates alternating current. The amount of induced current depends on the rate at which the rotor changes relative to the magnetic field. Because the magnetic field lines that pass through the coil are proportional to the cosine of the shaft angle, and the rate at which the field changes is the time derivative of the number of lines, the generated output waveform is a sine wave. One reason that sine waves are used in electrical circuits is because the sine wave results from this form of mechanical motion.

The output voltage from the rotating coil is an alternating voltage that, when connected to a load, will cause an alternating current or ac. If direct current is needed, there must be a method of changing this ac to dc. One very simple method of providing dc is to switch the polarity of the coil every half-rotation. Rather than use a single ring for connecting the moving coil to the external circuits, a segmented ring is used. This type of contact is called a *commutator.* The result is that every half-cycle the polarity of the connections to the rotating coil are reversed, and the result is a single polarity output or direct current, dc. However, the delivery of the energy to the dc circuits is not steady but occurs in pulses, which is undesirable.

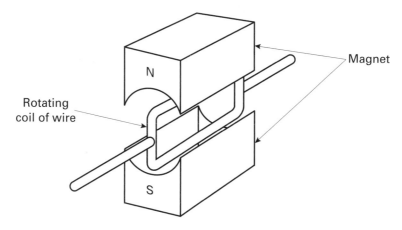

Figure 3-2   Simplified diagram of a dc generator.

What is needed is a smoother transfer of energy, and one method of achieving this is to increase the number of pulses, which can be done by increasing the number of rotating coils. We could use the same commutator arrangement to switch from one coil to another as well as to reverse the polarity and provide a large number of energy-transferring pulses.

Another method of providing an increased number of energy-transferring pulses is to increase the number of stationary magnetic fields. In the example of Fig. 3-2, the rotating coils pass through one fixed magnetic field per rotation. Two magnetic fields would cause the rotating coil to pass through two magnetic fields per rotation. If a generator had eight rotating coils and two fixed magnetic fields, 32 pulses of energy would be transferred each rotation, which begins to approach continuous energy transfer. Figure 3-3 shows a field arrangement providing 4 fixed magnetic fields.

For many years, dc generators used mechanical commutators to convert the basic ac generation to dc. The device was called a *dc generator* and had brushes and segmented commutators to make the ac-to-dc conversion. But brushes and commutators are a source of wear and failure, and this type of generator has been replaced with an ac generator called an *alternator*. The ac output of the alternator is converted to dc by using diode rectifiers. Figure 3-4(a) shows an alternating current generator using a half-wave rectifier to convert the ac output of the generator to direct current.

Although it does not have a segmented commutator, the alternator does have brushes and slip rings. There are also brushless alternators that have no moving electrical contacts, which produces a highly reliable alternator.

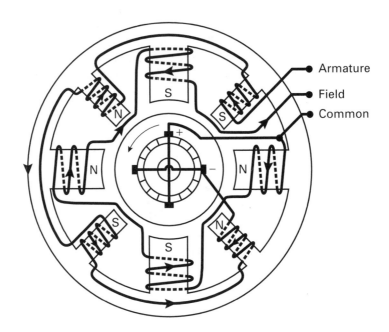

Figure 3-3 Diagram of a dc generator showing multiple poles.

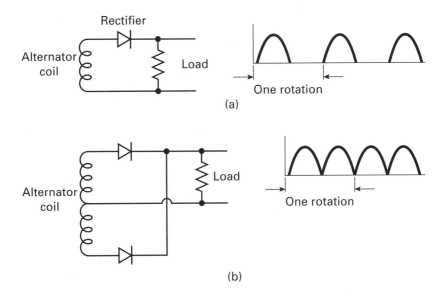

Figure 3-4 (a) Half-wave and (b) full-wave rectifiers connected to an alternator output.

The diode rectifier shown in Fig. 3-4(a) passes current in only one direction and thus converts the bidirectional output from the alternator to direct current. Although the rectified output is "direct current," it is far from steady current. There is a pulse of current that lasts for one half-cycle, which designates this circuit a *half-wave rectifier.* An improvement may be made by providing two outputs from the rotating coil by making a center tap for the coil. Now two diodes are used and each diode provides one pulse per cycle. This circuit is called a *full-wave rectifier.*

This is still far from a steady dc current and further improvement is desired. By providing three stationary coils separated by 120°, the generator provides three outputs; each is a sine wave and separated by 120°, as shown in Fig. 3-5. Each output is rectified using

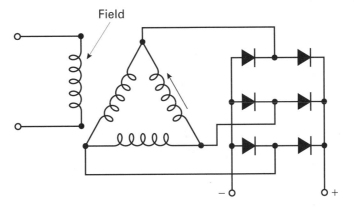

Figure 3-5 A three-phase alternator with a full-wave rectifier.

two diodes each; this results in six pulses per rotation of the generator shaft, which produces a reasonably steady dc output.

This form of generator is called a *three-phase alternator* (Figure 3-6), and variations of this are used to generate both dc and ac power. The example three-phase alternator consists of three stationary coils with one rotating magnetic field.

In an alternator, either the magnetic field or the pickup coils can be made to rotate; it is only necessary that the output coils and the magnetic field rotate relative to each other. In the case of the dc generator, the inherent ac nature of the output coils is converted to dc by the commutator. This requires that the magnetic field remain stationary. In the case of an alternator, the field can rotate and the output coils remain stationary. A current path must be provided to the field coils, which is done with simple slip rings.

The brushless alternator represents the greatest reliability of all alternator designs. Figure 3-7 shows a simplified diagram of a brushless alternator. The alternator has two stator coils. The first stator is an exciter stator, which provides the variable magnetic field for voltage regulation. The rotor that turns in the field of the exciter coils is a three-phase rotor with diodes, just as would be found in any alternator. However, rather than removing the energy generated by the rotating three-phase coils using slip rings, the three-phase output is rectified with six diodes in a full-wave rectifier, and the output current is used to generate a rotating magnetic field. This rotating magnetic field is surrounded with a three-phase stator that produces three-phase ac. This output can be used as a source of ac or rectified to produce dc for battery charging.

The brushless alternator generates an ac voltage in the rotor that is rectified to create a rotating dc magnetic field, which induces an ac voltage in fixed coils. The advantage of this alternator is that there are no brushes, but the disadvantage is that there are two stator coils and two rotor coils. The extra rotor and stator coils are sources of energy loss due to the resistance of the wire. These extra coils add to the weight and size of the alternator,

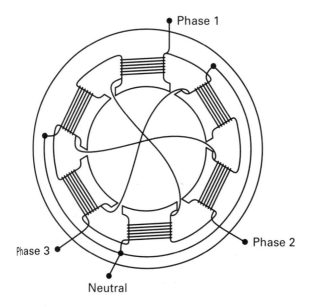

Figure 3-6   Example of a three-phase alternator.

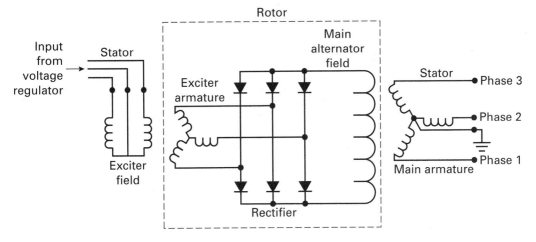

Figure 3-7   Diagram of the brushless alternator.

which is also a disadvantage. Finally, the diodes in the rotor add to the energy loss of the alternator. However, for the increased reliability obtained from the alternator, these losses are a small price to pay.

The alternator is clearly a superior method of generating electrical energy for the aircraft dc system and has become the predominant method of electrical power conversion. However, there is one application for which the older, commutated generator has an advantage. This application is the combination of motor and generator that serves as the starting motor for a gas turbine engine and then, after the engine is running, serves as the generator. In this application, one unit performs two functions, and thus the weight savings is worth the disadvantage of the mechanical brushes. Another advantage of the starter–generator is that the unit is permanently connected to the driving engine. The motor is energized to start the turbine, and after the turbine has started, the turbine powers the generator, which produces electrical power. The dc generator can serve as a motor, whereas there is no equivalent motor function for the alternator with current technology. Theoretically, such a device is possible, however.

Voltage Regulators

When generating dc, the frequency of the ac output of the alternator, which is directly related to the shaft speed, is not important. The output voltage, however, is important and is also directly related to the generator shaft speed. The output of the generator is used to charge a battery, and thus the voltage must be greater than the battery voltage to ensure that current flows into the battery. In an aircraft for which the generator is driven from the propulsion engine and there is a wide variation in shaft rpm, if some method is not provided to automatically adjust the generator or alternator output voltage, the battery will be either not charged or overcharged. The output voltage is proportional not only to the shaft speed, but also to the strength of the magnetic field. Therefore, the output voltage may be adjusted by providing a variable magnetic field.

Like the dc generator that for many years mechanically changed ac to dc, mechanically operated voltage regulators were used with the dc generators. The schematic of such a device is shown in Fig. 3-8. A special relay that opens its contacts when a specific voltage has been reached causes the short across the resistor to be removed and places the resistor in series with the field coils. This reduces the magnetic field and hence the output voltage from the generator. This causes the voltage-sensitive relay to no longer be able to hold its contacts open, and the short is replaced across the resistor and the output voltage from the generator begins to rise. This causes the relay to continually open and close as the generator output voltage oscillates around the point where the relay activates. Because the contacts open and close a large number of times, only small field currents can be handled without damaging the contacts. This type of voltage regulator is used only in small generating systems typically found on light aircraft.

An improvement to the voltage-sensitive relay is the addition of a second winding on the coil. The second winding passes the field current and is wound in the opposite direction to the voltage-sensing winding so that the passage of the field current counteracts the voltage winding. Increasing output voltage will eventually cause sufficient magnetic field to pull the relay's contacts open. This opening of the contacts interrupts the field current, which removes the counteracting magnetic field and further increases the attractive force, thus causing the contacts to rapidly open. This increases the speed of the voltage-sensing relay.

The use of a generator requires another relay, as shown in Fig. 3-8(a), to prevent reverse current from the battery flowing into the generator. This relay is a voltage-sensitive relay with a set point significantly lower than the voltage-regulator relay, but higher than the battery voltage. In addition to a voltage coil, the reverse current limiter has a current coil wound such that current in the reverse direction will counteract the magnetic field generated by the voltage coil and assure that the relay opens.

The source of current for the magnetic field is the battery. How is it, then, that if the reverse current relay is open and there is no battery current the generator will provide any output voltage to energize the reverse current relay? The generator field is constructed using steel, which has some residual magnetism. This magnetism permits the generator to provide an output voltage at very low levels of output current to energize the relay and make the battery voltage available for full generator operation.

The proper operation of relay-activated voltage regulators depends on the careful manufacture and adjustment of the mechanical relays. The older vibrating relays caused radio noise that had to be filtered and shielded, and the mechanical contacts pitted and wore. These mechanical units have been replaced with electronic regulators that are much more reliable.

One important difference between an older dc generator and an alternator are the diodes used to rectify the ac. These diodes provide an automatic reverse current protection, and thus a reverse current relay is not required for an alternator. Since most aircraft engines no longer have generators, most voltage regulators do not have reverse current relays.

Some newer voltage regulators substitute an accurate electronic voltage sensor, but maintain a relay to switch the field current, while more sophisticated regulators vary the

DC Power Systems

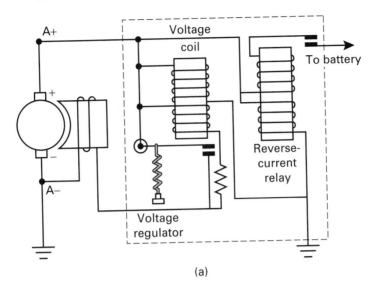

(a)

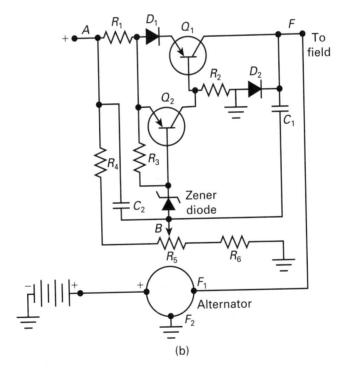

(b)

Figure 3-8 Example of (a) a two-relay voltage regulator and (b) an all-electronic voltage regulator.

field current by using transistors, thus eliminating mechanical contacts completely. This design represents the modern voltage regulator.

Figure 3-8(b) shows a modern, all electronic alternator voltage regulator. The output voltage is sensed and compared to a reference. This reference is a temperature-variable reference voltage to allow for the different battery potentials at various temperatures. The current to the field coils is passed through a transistor, which allows for a continuous variation of field current, rather than the on/off characteristics of the relay. Thus the output voltage is smooth and accurate, with the required compensation for temperature variations.

STATIC INVERTERS

In light aircraft, generating ac power is a very difficult procedure because of the requirement of constant-speed rotation of the ac generator. Small aircraft only have propulsion engines for a source of energy for electrical power generation, and this implies that an engine-driven generator would be powered from a variable-speed source.

For this reason, smaller aircraft have avoided ac power, opting for only dc-powered equipment. With the demise of the vacuum tube, the dc-only power source is no detriment and very sophisticated avionics systems are possible with dc power.

There are some areas where ac power is required. These systems typically involve indicators and autopilots and require the 400-Hz source for operating special motors called *servo motors*. Typically, these systems do not require a large amount of 400-Hz power because the servo systems operate light structures such as indicators. Mostly, the 400 Hz is used as a reference for position sensing and other control functions.

Although not much power is required of the 400-Hz source, the servo motors work best when the 400-Hz source is clean, free of noise, and stable in frequency and voltage. Often, this 400-Hz source is generated with a device called a *static inverter*, which is a 400-Hz power supply that changes 28 V dc to a constant frequency and amplitude 400-Hz source. The term *static* distinguishes the inverter from a mechanical device called a *rotary converter*, which is a constant-speed dc motor turning an ac generator to provide a 400-Hz output. This device is inefficient and heavy and is rapidly being replaced by all-electronic static inverters.

A block diagram of a 400-Hz static inverter is shown in Fig. 3-9. The 28-V dc input is converted to a sine wave with an oscillator and fed to an amplifier that drives the output transformer. The oscillator is an electronic circuit that provides a very stable 400-Hz signal, which is unaffected by variations in the 28-V supply voltage. The output voltage is sampled by rectifying the output and comparing the rectified voltage to a dc reference voltage. The difference between the rectified ac voltage and the reference voltage is used to control the amplitude of the voltage driving the power amplifier. This combination provides a constant-frequency and constant-voltage 400-Hz source. Typical static inverters provide power outputs from 10 to 1000 W.

Because the 400-Hz power supply is used by a number of critical systems, two static inverters are often employed, with one inverter serving as a backup. In addition, a 400-Hz distribution system with buses and circuit breakers is provided for protection of the 400-Hz power system and so that faults can be isolated.

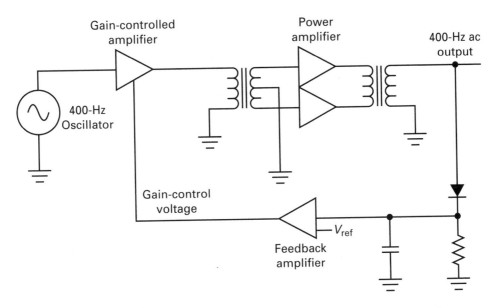

Figure 3-9 Block diagram of a static inverter for generating a constant amplitude and frequency 400-Hz source.

WIRES AND CABLES

Electrical current is conducted through wires of varying sizes. A number of important characteristics of wires should be understood to fully appreciate electrical distribution systems.

There are two basic components of electrical wire, the conductor and the insulator. And wire can be divided into two groups, solid and stranded. The stranded wire is made of a bundle of smaller wires, while solid wire is one solid conductor. The stranded wire is more flexible and resists fracturing if it is flexed. The advantage of solid wire is that it is less expensive and there are some applications for which a flexible wire is not desired. For power distribution in an aircraft, no solid wire is used. This is in contrast to power distribution in buildings, where solid wire is used almost everywhere. Aircraft use of solid wire is in motors, relays, and solenoids, where the wire is wound on a frame and cemented in place. The rigidity of the solid wire helps keep the wire in place.

The insulation of electrical wire prevents the circuits carried by the conductor from shorting to other circuits or the airframe. A variety of insulation types is used in aircraft wire, but most are plastic coatings. A common form of wire insulation, polyvinyl chloride, is used only sparingly in aircraft. This is because it has an undesirable characteristic; it burns and gives off a toxic gas. Another popular insulation, Teflon, does not burn under normal circumstances, but has its own undesirable characteristic, cold flow. If a Teflon-insulated wire is supported by a sharp edge, the edge will slowly penetrate the insulation.

A number of other insulation materials are used for aircraft wire, such as fluorinated ethylene propolyne, polymide, and Kapton. Each material has special characteristics, such as abrasion resistance, chemical resistance, sunlight resistance, (ultraviolet), and the like.

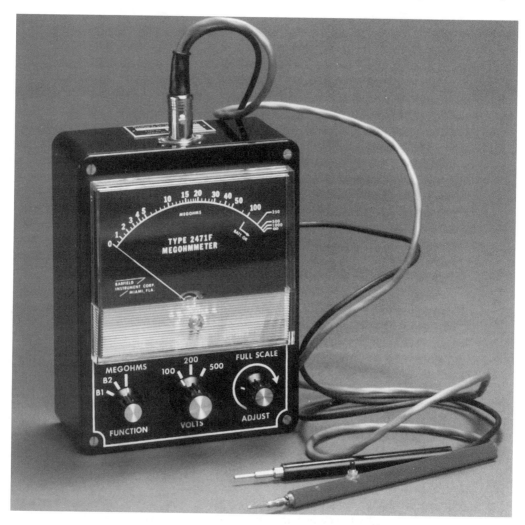

Photo 3-1   The Barfield 2471 megohmmeter is used for measuring very high resistances to test aircraft wiring insulation. Photo courtesy of Barfield, Inc.

Various modifications of plastic-insulated wire enhance the characteristics of the insulation. For example, by providing a woven glass fiber jacket, the abrasion resistance of any wire can be improved dramatically. On to increase abrasion resistance and reduce the cold flow tendencies of some plastic insulations, the wire may be exposed to gamma radiation, which hardens the outside of the insulation jacket. This treatment gains the advantages of high temperature capability and chemical stability without the softness of some plastics.

The thickness and material of the insulation determine how much voltage the wire may have present before the insulation breaks down and allows current to flow from the wire to its surroundings. Most wire can handle from 500 to 2000 V safely, which is suitable for most aircraft applications. The breakdown capability of a wire is a peak voltage, not an rms value. Therefore, for a 110-V ac system the peak voltage is 157 V. Some aircraft electrical devices cause large transient voltage excursions that can exceed 600 V even on 28-V dc lines. The breakdown of a wire is a destructive event that causes gradual weakening of a wire.

Wire is manufactured in a variety of standard sizes. Unfortunately, wire sizes are designated by a rather arbitrary system called the American Wire Gauge (AWG). An example of some wire sizes is shown in Fig. 3-10. The AWG system is unfortunate only in the fact that there is no easy relationship between diameter, area, resistance, or any other physical parameter of wire and the appropriate AWG size. The higher the AWG number is, the smaller the wire. As an example, number 40 AWG is a hair-thin wire used for winding magnets and coils of delicate instruments. On the other end of the scale are the numbers zero, 0, double zero, 00, and triple and quadruple zero, 000, 0000. The diameter of the conductor of 4/0 wire is almost ½ inch.

Every student usually wonders how much current you can pass through a certain size wire. The amount of current depends on a number of factors. To understand these factors, interconnecting wires must be looked at as resistors. As good a conductor as copper is, it still has some resistance, and it is the resistance that will limit the amount of current.

Because of this resistance, there will be some voltage drop between the source of energy and the load. Referring to the wire table, if a 28-V device that draws 10 A is connected to the power source using 20 ft of number 18 wire, there would be a voltage drop of

$$10 \text{ amps} \times 6.39 \times 10^{-3} \ \Omega/\text{ft} \times 20 \text{ ft} = 1.28 \text{ V}$$

This represents a 5% drop of voltage, which may not be desirable. This calculation involves the wire from the 28-V supply to the equipment. There is a return path through the

AWG Number	Diameter/inches	Ohms/1000 feet
0000	0.46	0.049
000	0.41	0.062
00	0.37	0.078
0	0.33	0.098
1	0.29	0.124
. . . .	. . . .	. . . .
12	0.081	1.59
14	0.064	2.53
16	0.051	4.02
18	0.040	6.39
20	0.032	10.15
22	0.025	16.14
24	0.021	25.67
26	0.016	40.81
28	0.013	64.90
30	0.010	103.2

Figure 3-10   Examples of some AWG wire parameters.

aircraft fuselage, which, in most electronic systems, is, ironically, called ground. Generally, the resistance of the return path through the fuselage is much lower than in the supply line. The voltage drop through the fuselage is less than the copper supply wire because, although copper is a much better conductor than aluminum, there is so much more aluminum present. This is the equivalent of a very large aluminum wire.

In spite of the fact that the return path has involved the metallic fuselage for many years, in some situations this is becoming a problem. Composites are being increasingly employed in the construction of aircraft. Although the majority of an aircraft is still aluminum, there are large panels of nonconducting composites. Should an electronic system have one of these large expanses of composite material between the system and its energy source, the return path resistance could increase dramatically.

In other situations, because the fuselage return path is shared by other systems, interference can occur. This phenomenon is called ground loops.

The 5% voltage drop in our example does not appear to be a serious problem because many systems can operate with 95% of their rated voltage. A larger-sized wire can reduce the voltage drop, but there are important advantages to using the smallest-sized wire practical. Small wire is less expensive, requires smaller connectors, and is lighter. A significant danger to using wire that is too small is the heating due to the energy dissipated in the wire's resistance. In our case the total power dissipated is 12.9 W.

This power is distributed over a 20-ft wire, which results in about 0.645 W/ft. The heat is generated in the copper, which is surrounded by the insulation of the wire, which is a good thermal insulator too. Other wires may surround the wire in question, and each wire may dissipate a similar power. Now place all this in a hot aircraft engine compartment, and the temperature may not melt copper, but it will certainly have some effect on common plastic insulations.

Maximum current-carrying capacities of copper wire are set as a function of the maximum permitted temperature rise and the maximum voltage drop. Wire sizes for long runs are limited by the voltage drop, whereas the shorter runs are limited by heating. Choosing a wire size that is too small will cause unnecessary energy waste due to heating, and choosing a wire size that is too large will add unnecessary weight to the aircraft and waste its fuel energy.

Power Distribution

It is necessary to distribute the power to the devices in the aircraft that require it. It may seem, on the surface, to be a trivial task to connect each item requiring power to the battery or generator. What is wrong with this simplistic approach is the protection against failures. In smaller aircraft, there is only one generator and essentially one source of power. In a larger aircraft, there can be as many as four engine-driven generators, plus an APU, as well as both dc and ac systems. To be reliable, the power distribution system must be capable of surviving failures of electrical equipment, including generators, without bringing down the entire power system. Total loss of electrical power in an aircraft during flight is about as catastrophic as situations become.

The technique for creating a reliable power distribution system in an aircraft is to divide the power system into sections so that each section can be isolated from the power distribution system should a serious failure occur. Figure 3-11 shows a simplified power distribution system for 28 V dc for a medium-sized executive twin aircraft. The analysis of this system could progress from the source to the users or from users to source, which will be used here.

Each user, such as avionics equipment, motors, and lamps, has an individual circuit breaker. These circuit breakers are not only overcurrent protection but serve as on/off switches. Typically, they are mounted on a panel with a number of other breakers. The individual circuit breakers allow one user system to fail and only that system will be removed from the power source, thus causing the least amount of disruption.

The supply side of a circuit breaker usually connects to a distribution form called a *bus.* Typically, these buses are not wire but rectangular strips of copper called bus bars. Since the bus bars are not insulated, they are contained in an enclosed area called a distribution box. Typically, the breakers are connected directly to the bus bars without intervening wires.

Bus bars are protected with circuit breakers, which protect against short circuits that occur before an individual user circuit breaker. One important function of a bus is to separate parts of the distribution system so that, if a bus failure should occur, only a fraction of the aircraft electrical system will fail. In the example of Fig. 3-13, three buses are shown. The first bus is the electrical equipment bus, which provides power to nonelectronic-type equipment such as gear motors, flap motors, cooling blowers, and lighting.

Bus systems are divided in a number of ways, but some of the more prevalent are the essential bus, which contains items that are required for safe operation, the emergency bus, which contains items that must be operational in an emergency situation, and other buses that can be separated by equipment type, such as avionics, motors, and lights. The separa-

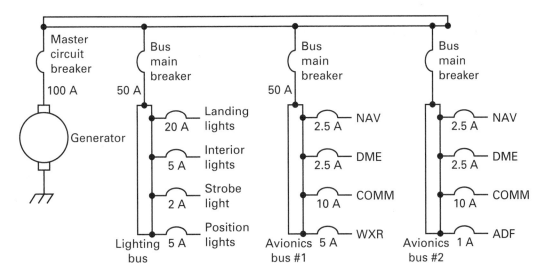

Figure 3-11   Simplified power distribution system.

tion of buses by equipment type is an aid in finding a breaker if a particular type of equipment must be switched on or off.

Also, in Fig. 3-11 there are two avionics buses that power the electronic equipment such as navigation radios, displays, and flight control system. The avionics are divided so that not all the navigation radios or communications radios are connected to one bus. Therefore, if an aircraft experiences an in-flight bus failure, the aircraft does not lose all radio navigation or all radio communication. Some systems are not redundant, such as the flight control system, and these systems are divided to equally distribute the load between the two bus systems.

Aircraft do not have large storage batteries because of the tremendous weight of large batteries. In a light aircraft such as the example at hand, the battery is needed to power the aircraft electronics for a few minutes until the engines are started and provide the necessary energy. The lifetime of a battery with all avionics systems operating would typically be less than an hour. To perform maintenance on a light aircraft, the aircraft is powered from a ground power unit, or GPU. For this example executive twin, the GPU would provide 28 V dc in lieu of the battery. This is not to charge the battery. Figure 3-13 shows provisions for providing a GPU. A large plug is provided in the aircraft fuselage where the GPU may be connected. Voltage from the GPU activates a large contactor, which is a special type of relay, that disconnects the battery and places the GPU on the 28-V line. The contacts of the GPU plug are staggered so that the actual connection of the ground power supply is made through the contactor, rather than the GPU plug, and prevents arcing and pitting of the connector.

In a twin-engined aircraft, there is usually one alternator per engine. During normal conditions, both alternators charge the aircraft battery and provide power for the equipment.

The generators typically provide individual buses for very heavy load devices, such as the gear motor and air-conditioning units. This prevents the noise and transients generated by these very heavy loads from affecting more sensitive equipment such as avionics.

A technique called *split bus* is used in large aircraft. This provides two separate bus power distribution systems, each with its own alternator and battery, and permits the two buses to be tied together in case there is an alternator failure. Figure 3-12 shows an example of this type of bus system. There are multiple equipment buses, but there is a new bus called the *tie bus*. In normal operation the tie bus joins the individual buses together through contactors called bus tie breakers. When a failure occurs, the failed alternator is removed by opening the bus tie breaker. This leaves the failed bus and its generator isolated, and the cause of the failure may be analyzed. If the failure is due to a defective alternator rather than a short, the alternator may be isolated by opening the alternator circuit breaker and reconnecting the bus to the tie bus. Generally, this requires that the electrical load of the aircraft be reduced because two generators are now supplying the load normally supplied by three.

The procedures outlined for isolating failures, switching buses, shedding loads, and so on, must be done with care. If these actions are not taken carefully and two buses are tied together incorrectly, the failure present on one bus can cause damage on the good bus. Printed emergency procedures should be followed when isolating electrical failures in an aircraft.

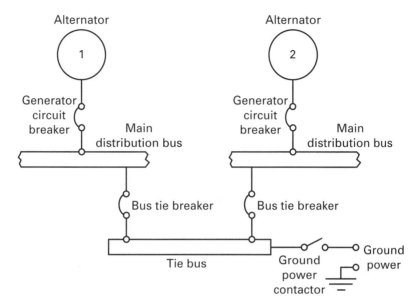

Figure 3-12 Simplified diagram of a split bus power distribution system.

Connectors

There are two basic types of aircraft connecting devices, plugs and jacks and terminal boards. Plugs and jacks are devices that allow equipment to be easily connected and disconnected, usually without tools. An important class of device that is fitted with connectors is called line replaceable units (LRUs). These units are often slid into mounting trays with connectors mounted at the rear of the tray that automatically engage when the unit is slid into the tray. In some cases these connectors are keyed so that only the correct unit may be installed in the tray.

Terminal boards are provided for equipment that may need to be rewired, but not often. Connecting or disconnecting wires on a terminal board usually requires simple hand tools.

For many years the only acceptable method of connecting wires to connectors and joining wires together was soldering. Significant problems are associated with soldering. First, it requires special tools. In a factory environment, soldering may be an acceptable technique, but when changes are made in the field, the use of a soldering iron may be difficult. Second, a good solder joint requires the right tools at the right temperature, the correct solder, good surface preparation, and cleaning of the joint to remove corrosive fluxes. A well-trained technician is needed to satisfy these requirements, because compromising on any one of them will produce a less reliable joint.

One disadvantage of soldering that no technician can avoid is the wicking of the solder up a stranded wire. This effectively causes a short length of the stranded wire to become solid. Unfortunately, this is also an area where the greatest amount of stresses can occur and result in a fractured wire.

Most modern connectors use wire crimping methods. As an example, to install a wire in a connector, the wire is stripped and inserted into a connector pin, and the connector pin is squeezed in a crimping tool. Not only is the conductor crimped tightly, but some connectors also crimp some of the insulation so that the wire is prevented from flexing at the connector pin. The connector pin is inserted into the connector with the wire attached.

Exhaustive studies have been made of crimping reliability, and it has been shown that modern crimping techniques are significantly more reliable than soldering.

AIRCRAFT LIGHTING SYSTEMS

Aircraft lighting is an important part of aircraft design. A number of light systems are used in a modern aircraft; they can be separated into interior and exterior systems.

Exterior lighting includes the position lights, landing lights, and anticollision lights. Interior lighting includes instrument panel lights, interior illumination lights, and emergency lighting systems.

The position lights or navigation lights are provided so that other aircraft can visually determine the approximate size and direction of travel of an aircraft at night. There are three position lights: a red light on the left wing tip, a green light on the right wing tip, and a white light on the tip of the tail.

The viewing angles of the lights are restricted so that not all the lights are visible in all quadrants, as shown in Fig. 3-13. With the lights as shown, the green and red are visible directly ahead of the aircraft, the green and white are visible to the left of the aircraft, and the red and white to the right.

An anticollision beacon is required of all aircraft of recent manufacture and of all aircraft engaged in night flight. Two types of anticollision lights are in use: the rotating beacon and the high-intensity strobe light.

The rotating beacon is a red lamp surrounded by a rotating mirror, which gives the effect of a blinking light. The strobe light is a more recent alternative and produces a blink-

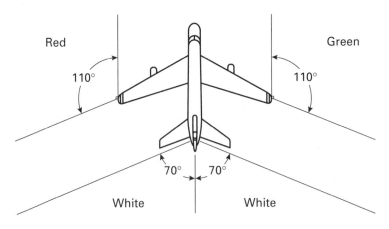

Figure 3-13   Field of view of the position and anticollision lights.

ing high-intensity light without moving parts. The strobe light uses a xenon flash lamp, the same type of lamp used in a camera flash lamp. A block diagram of the strobe light is shown in Fig. 3-14. A quartz tube is filled with xenon gas that is subjected to a high electric field. The electric field is slightly less than that which would cause ionization. A second electric field is generated with a short pulse of voltage such that the resulting electric field does cause ionization. This high-voltage pulse is created with a transformer and starts the ionization. Once the gas ionizes, the gas becomes a conductor, and the energy storage capacitor is discharged through the xenon, which gives off an intense pulse of light. The energy for the pulse of light is stored in the capacitor, and the ionizing gas causes a heavy current flow that discharges the capacitor to a voltage level that cannot support the current flow, whereafter the current ceases. Immediately, the capacitor begins to charge and the next light pulse can be triggered.

The typical duration of the light pulse is 0.001 s and peak powers of 1000 W are common. The strobe type of anticollision light is very effective and has a range considerably longer than the conventional rotating beacon.

Example

What is the peak power of a 0.001-s light pulse generated from a xenon flash tube when the efficiency of the tube is 60%? The capacitor is 10 microfarads (μF) and is charged to 450 V.

Solution.   The energy stored in the capacitor is $CV^2/2$ or

$$\text{energy} = \frac{10^{-6}}{2} \times (450)^2 = 1\text{J}$$

Since 0.01 J of energy is applied to the flash tube in 0.001 s, the power during that period of time is

$$\text{electrical power} = \frac{\text{energy}}{\text{time}} = \frac{1\text{ J}}{0.001\text{ s}} = 1000\text{ W}$$

Landing lights provide illumination forward of the aircraft for night landing. These lamps are usually sealed beam spotlamps similar to automotive headlights.

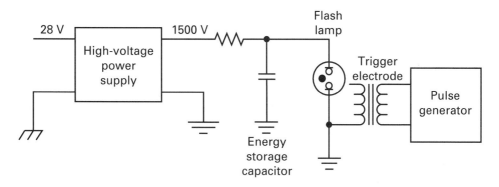

Figure 3-14   Block diagram of a strobe-type anticollision light.

Interior lighting includes the usual interior reading and floor lighting for passengers. Emergency exit lighting is provided for larger aircraft. Also provided is an emergency floor lighting showing the location of and the path to emergency exits. These emergency lights are powered in an emergency situation by internal dedicated batteries that provide this lighting even if every bus system in the aircraft, including the emergency bus, has failed.

An important lighting system is the instrument lighting. This system provides variable-intensity lighting for the instruments located in the instrument panel. The instrument panel is often divided into several sections so that each section can be separately adjusted for lighting intensity. Each navigation or communications radio, instrument, weather radar, Electronic Flight Instrumentation System (EFIS), and so on, must be capable of being reduced in intensity so that one indicator does not overwhelm the others.

A lighting bus provides a standard voltage for setting the intensity of panel-mounted instruments so that dimming the panel does not require that each piece of equipment have a brightness control that must be adjusted.

There are three lighting bus voltage standards: 0 to 5 V dc, 0 to 5 V ac, and 0 to 28 V dc. The standard of brightness is the intensity that would result from connecting an incandescent lamp to the lighting bus. As an example, if a 0 to 5 V ac bus were in use, an indicator that does not use incandescent lighting should match the brightness of a 5-V incandescent bulb connected to the lighting bus. As an example, a weather radar display will provide the dimming of the CRT by adjusting the electron beam intensity. Circuits within the weather radar assure that the intensity of the weather radar matches that of a 5-V incandescent bulb connected to the 0 to 5-V lighting bus.

REVIEW QUESTIONS

3.1. Name some of the sources of electrical energy used in aircraft operation.

3.2. What restrictions are there relative to shaft speed for generating 400-Hz ac power?

3.3. What are the advantages of multiphase power generation and distribution?

3.4. What is a primary cell? What is a secondary cell?

3.5. Why must batteries be vented?

3.6. If the avionics of an aircraft requires 28 V at 14 A, how long can the avionics be operated from a 72-A-h battery before the battery is discharged?

3.7. What limits the maximum current that may be handled by a wire in an aircraft?

3.8. Why are electrical power buses used in an aircraft?

3.9. How much power per foot will a number 12 wire dissipate if it is supplying 12 A?

3.10. Why are alternators used rather than generators when generating dc for battery charging?

3.11. What functions do voltage regulators for dc generating systems perform?

3.12. What color are navigation lights and where are they mounted?

3.13. What is an anticollision beacon?

3.14. Why is a static inverter used?

3.15. What unique characteristic does the lighting bus have as compared to other power buses?

4

Communications Systems

CHAPTER OBJECTIVES

In this chapter the student will learn the basic physics of radio wave propagation and the effects of natural and human-made noise. The student will study the basic construction of transmitters and receivers, with special emphasis on the superheterodyne receiver and the role of frequency synthesizers. An understanding of interaircraft communications will be gained from this chapter. The construction and characteristics of microphones, particularly those used in airborne communications, are covered.

INTRODUCTION

Communications was the first requirement for controlling aircraft traffic. Requesting and granting permission to land, assigning cruising altitudes, and other necessary communications were the driving force behind providing aircraft with communications equipment. In the case of military aircraft, the installation of communications capability in the aircraft marked the difference between a fighting tool and a curiosity.

To gain historical perspective, we note that broadcasting started in the United States in 1920. One major hurdle in establishing a broadcast industry was inventing radio transmission equipment capable of transmitting voice and music; Morse code transmissions had taken place for some time. Designing equipment small and light enough to transmit and receive voice from an aircraft was a major undertaking. In fact, some early attempts at airborne communications actually used Morse code.

The great popularity of radio broadcasting brought enormous strides in radio transmitter and receiver technology. Smaller, more effective, and more reliable vacuum tubes permitted lighter receivers and transmitters to be manufactured. Perhaps the biggest advance in radio receivers took place in 1920 when a patent was issued to Edwin H. Armstrong for the superheterodyne receiver. This receiver topology provided solutions for some very difficult problems associated with the radio receivers of the day.

RADIO COMMUNICATIONS

Communications is the process of transmitting information from one location to another. A communications system requires a communications medium for the conveyance of the information to be exchanged. For telephone and cable television systems, the medium is cable and fiber optics. For the case of aircraft, it is clear that the communications medium must be electromagnetic fields or radio waves, because cables and fiber optics are impractical for airborne communications.

When we studied the transformer, we saw how energy could be transferred from the primary to the secondary through the alternating magnetic field generated by the primary. The transformer was made so that the maximum flux lines were coupled and the transformer would be efficient. But, rather than close coupling the coils, what if the coils were placed at a distance? Could there be energy transfer? The magnetic lines of force from a solenoid extend, theoretically, to infinity. Although there is considerable dilution of the energy, there should be some energy pickup at any distance. Remember that the transformer action only occurred with an alternating magnetic field from the primary. It can be deduced from this approach that an alternating magnetic field is capable of propagating energy over a distance.

As to energy transfer by an electric field, a law of physics states that every alternating magnetic field is accompanied by an alternating electric field. It does not matter which field is the generated field; the creation of one field automatically generates the other. The magnetic lines of force and the electric field vectors are perpendicular to each other and radiate away from the source in straight lines. The direction of propagation of energy is perpendicular to both the electric field and magnetic field vectors. Remember, only changing fields propagate and are capable of transporting energy. The most common changing electric or magnetic field is generated by a sinusoid alternating current, which produces a sinusoid electromagnetic field, or radio waves. These are the "Hertzian" waves that Gugliermo Marconi demonstrated by spanning the Atlantic Ocean in 1902 with radio signals, one year before the Wright Brothers flew.

Example

Radio transmitters operating at low frequencies used for navigation purposes transmit a vertically polarized electric field. This field is generated by using a vertical tower. What is the orientation of the magnetic field?

Solution. A magnetic field vector that is perpendicular to a vertical electric field must be parallel to the ground. However, it must also be perpendicular to the direction of radiation. To radiate away from a vertical tower implies that the direction of radiation is like the spokes in a wheel. Therefore, for the magnetic field vector to be parallel to the ground as well as perpendicular to the spokes, it must be tangential to a circle with its center at the radiating tower.

Electromagnetic waves are used to transmit information by wave motion. As covered briefly in Chapter 2, an alternating electric field is necessary for energy propagation, and a sine function would be the obvious choice for the alternating electric field. A sine function conveys no information. What is needed is a method of adding the information to the sine function. The equation for an electric field that is proportional to a sine function with time t as the independent variable is

$$E(t) = A \sin (2\pi ft) \tag{4-1}$$

where A is the amplitude of the electric field, and f is the frequency of the sine function.

Applying the information to be transmitted to the sine wave requires that the wave, which is called a *carrier* because it carries the information, be changed in some fashion as a function of the information. Only two parameters of Eq. (4-1) can be changed to apply the information, the amplitude A and the frequency f. When the information is applied to the carrier by modification of the amplitude, the process is called *amplitude modulation.* When the frequency is modified by the information, the process is commonly called *frequency modulation.* The term $2\pi ft$, which contains the frequency, is actually an angle, and the modification of the frequency to modulate the carrier with the information to be transmitted is correctly called *angle modulation.*

Radio-wave Propagation

One very important characteristic of radio waves is that energy propagation can take place without a physical medium. In other words, radio waves can propagate through an absolute vacuum. This is different than sound waves. Sound energy must travel through a physical medium such as air or water. This concept was very controversial around the time Marconi began transmitting radio signals. Many scientists believed that there must be some substance even in the dark nothingness of outer space. This substance was called ether. Albert Einstein postulated and later proved that there was no ether and that electromagnetic energy requires no propagation medium.

If radio waves propagated in space with no objects present, whether they be as small as molecules of air or as large as a planet, the direction of propagation would be away from the transmitter in straight lines. However, objects affect straight-line propagation in a number of ways. Large objects reflect or completely stop the radio-wave propagation, while very small objects such as molecules have a very small effect on propagation. On the other hand, there is a huge number of molecules, and even though each molecule has only a small effect on propagation, the net result from the large number of molecules can be very pronounced.

One effect that is caused by large reflecting objects is called *multipath.* This phenomenon occurs when a radio wave has traversed two or more paths from a transmitter to a receiver. When the radio wave arrives at the receiver, the two received waves do not have peaks and valleys that occur together because of the two different paths, sometimes adding in phase and enhancing the wave and at other times adding out of phase and canceling the wave. The result is a distortion of the received wave. This phenomenon is a very common problem with radio transmission and is called *multipath distortion.*

All radio waves propagate at the same speed, 3×10^8 meters per second (m/s). Let us imagine for a moment that you could take a stop action photo of a radio wave. If we were to photograph a transmitting tower, we would see the radio waves with their motion frozen in time. We would see two waves; the electric field wave and the magnetic field wave, each having the same wave length but always perpendicular to each other. We would also see that the phase angle between the magnetic wave and the electric field was 90°. That is, the electric field is crossing zero when the magnetic field is at a maximum. The reverse is also true; when the electric field is at a maximum, the magnetic field is crossing zero. The electric and magnetic field waves represent the transmission of energy. Because of the 90° relationship between the electric and magnetic fields, when the electric field is zero, which indicates that the electric field contains no energy, the magnetic field is at a maximum, indicating that the magnetic field contains all the propagating energy.

The waves result from the electric and magnetic fields completing a cycle. As each cycle occurs, one radio wave is generated, which moves away from the tower and is captured by our photograph. How long would each wave be? It would depend on how long it took to complete one cycle: the longer it takes, the longer the wavelength. Therefore, one wavelength is

$$\text{wavelength} = C * T_0 = \frac{C}{f} \tag{4-2}$$

where C is the velocity of electromagnetic radiation, T_0 is the period of one cycle, and f is the frequency of the radiation.

Example

What is the wavelength of a standard broadcast station operating at a frequency of 1000 kilohertz (kHz)?

Solution. The wavelength is $3 \times 10^8/1 \times 10^6 = 300$ m.

Not all electromagnetic waves are radio waves. Although all electromagnetic waves have the same characteristics, such as wavelength, intensity, and straight-line propagation in space, electromagnetic waves of differing wavelength interact differently with various materials. As an example, a window shade will block out electromagnetic light energy and make a room dark. A radio in the same room, darkened with shades, is not affected. It is clear that the radio waves pass through the shades and the walls of the room easily.

There is significant variation in the wavelengths of radio waves used for aircraft purposes, and even though they are all radio waves, they react differently to their environment. The choice of a wavelength for an aviation application depends on the characteristics required for that application.

FREQUENCY SPECTRUM

There are a large number of users of radio communications. How is it that these users can coexist without interfering with each other? Radio communicators can operate without interfering by choosing different radio frequencies. A radio-frequency carrier is needed to generate the required electromagnetic waves. The information is applied to the carrier by modulation. But each modulated carrier has a specific frequency that can be used to distinguish one transmission from another. Communications by radio that take place at one frequency do not interfere with communications on another frequency. This is because the radio communications receiver can be made to filter out from all possible frequencies only that frequency where the desired communications is taking place.

Everyone is familiar with standard broadcast radio. Each broadcast station is assigned to operate at a specific frequency. In the case of the broadcast band of frequencies, there are broadcast stations operating at every 10 kHz from 540 to 1700 kHz. These 10-kHz frequency assignments are called *channels*. More than one user may be assigned to a channel, but the users that share a channel are separated by a distance such that they will not mutually interfere.

Describing the distribution of radio energy as a function of frequency describes the radio *spectrum*. One example of a spectrum would be a plot of the strength of radio stations operating in the broadcast band as received at a specific location.

Bandwidth

The information to be transmitted has a specific bandwidth. Basically, if more information is transmitted per second, more frequency spectrum is required to pass that information. This is because of the natural physical relationship between time and frequency. As an example, the sound waves that comprise normal speech range from a low frequency of 300 hertz (Hz) to a high frequency of 3 kHz. The bandwidth of speech is 2.7 kHz, which is the total frequency range of speech. When this information is transmitted by radio, a range of frequencies of no less than 2.7 kHz is required to reliably transmit speech information. This is the theoretical minimum. Typically, more bandwidth is required to pass the speech information, usually more than twice the bandwidth of the speech information. When allocating a part of the radio spectrum for communications, each radio transmitter must be given a band of frequencies wide enough to pass the required information.

Radio-wave Propagation

If we were to generate our radio waves in outer space, they would propagate in precise straight lines. However, on Earth a number of effects cause radio waves to take paths other than straight lines. We attempt to take advantage of these effects, but at other times the lack of straight-line propagation causes difficulties.

There are two significant characteristics on Earth that cause radio waves to take other than straight-line paths. First, Earth is a conductor of electricity. The ability of Earth to conduct is mainly a low-frequency phenomenon occurring below about 3 megahertz (MHz). When low-frequency electromagnetic waves, particularly vertically polarized, propagate along the surface of the earth, the waves cause currents to be generated in the soil, just as the magnetic field causes current to be induced in the secondary of a transformer. This current flow in the earth causes its own magnetic field, which bends the propagating wave and causes it to stay close to the earth's surface. This kind of propagation is called a *ground wave.* Although the ground does some conducting, it is not a good conductor and has considerable resistance. Because of this resistance, a lot of energy is lost to heating the ground and ground-wave signals tend to dissipate rapidly. This loss is simply overcome by providing large amounts of transmitter power to offset this loss. Clear channel radio broadcast stations, as an example, operate with transmitter output powers of 50,000 watts.

Earth is surrounded by another conductor, the ionosphere. This is a layer of ionized gases at the upper reaches of the atmosphere that can act as a reflector of radio waves. Radio waves that would be headed for outer space are turned back to Earth at great distances from the transmitter. Only a relatively small amount of energy is lost in the reflection, and the signal loss for very long distance transmissions can be small.

Not all radio waves are reflected by the ionosphere. If the frequency is too low or too high, it passes through the ionosphere to be lost forever to outer space.

The amount of ionization in the atmosphere determines what these lower and upper frequency limits to ionospheric propagation will be. Radiation from the sun is responsible for the ionization of the outer atmosphere. Therefore, the conditions of the sun affect radio propagation. First, the time of day has a lot to do with ionospheric propagation. Clearly, there will be more ionization during the day than at night. This does not imply, however, that there can be no ionospheric propagation at night, only that the propagation at night will be different than during the day. The distance from Earth to the ionosphere also changes between day and night, which has a significant effect on propagation.

Another attribute of the sun that has an effect on the propagation of radio waves is the number of sunspots, areas of intense electromagnetic activity on the sun's surface. This activity causes solar radiation to increase, which causes more ionization in the ionosphere. The number of sunspots and consequently the level of electromagnetic activity change over an 11-year cycle. Therefore, there are 5.5 years of good ionospheric radio propagation and 5.5 years of poor propagation.

There is a maximum frequency above which radio waves pass through the ionosphere, which is mainly a function of the time of day and the sunspot cycle. It is called the *maximum usable frequency* or MUF. Basically, frequencies above the MUF will pass right through the ionosphere and will propagate into space, whereas frequencies below the MUF will experience some ionospheric propagation. Using a frequency below the MUF does not ensure that the signals transmitted will propagate to the area of the world desired, only that ionospheric propagation will take place.

Because of these variations of the ionosphere, radio frequencies must be chosen as a function of the 11-year sunspot cycle, the time of day, and the distance to be covered. The necessary range of frequencies to provide needed communications is significant. It is not a matter of changing an operating frequency by 5% or 10% but may require using twice the frequency. Therefore, high-frequency radio communications equipment covers a broad range of frequencies.

In the aircraft VHF communication band of frequencies, from 118 to about 136 MHz, there is practically no ionospheric propagation. Without the effect of ground wave and ionospheric propagation, radio waves would propagate in straight lines. However, radio line of sight, which is the maximum possible radio propagation distance without the help of the ionosphere or ground waves, is greater than the geometric horizon.

There are two basic propagation mechanisms for extending VHF radio waves beyond the visual horizon. The mechanism that provides most of the propagation beyond the visual horizon is *refraction.* When electromagnetic energy passes from one index of refraction to another, the path of the electromagnetic energy is deflected unless the electromagnetic energy is perpendicular to the boundary between the two indexes of refraction. Radio waves travel through air, which has an index of refraction slightly greater than but very close to 1. The index of refraction of a vacuum is exactly 1. The index of refraction of the atmosphere varies from slightly greater than 1 to exactly 1 from the surface of Earth to the vacuum of outer space.

When radio waves go from a region of higher index of refraction to an area of lower index of refraction, the energy is deflected away from the normal, which is the direction perpendicular to the boundary. Therefore, radio waves launched at small angles to the surface of Earth will be deflected back toward Earth and thus can go beyond the visual horizon, as shown in Fig. 4-1. The amount of deflection is small because the change in the index of refraction of air is small, and thus the radio range is typically only 15% greater than the visual horizon.

The second mechanism for extending radio range beyond the geometric horizon is *scattering.* Radio waves are reflected from air, dust, and water in a range of directions. Some of the reflection is in the direction of the intended receiver, and this phenomenon extends the radio range beyond both the geometrical and radio line of sight. Only a small percentage of the energy that is reflected is scattered in the direction of the receiver over the horizon, and thus the signal diminishes very rapidly beyond the radio line of sight.

Everyone has seen an example of both refraction and scattering. When the sun sets, the image of the sun can become a large red disk, which is due to the refraction of the sun's light from below the geometrical horizon. Although the image is distorted, the shape of the sun is clear. Once the red disk has disappeared, the orange-red glow on the horizon is from scattering. The sun is no longer a distinct disk, but the light is still visible, although greatly reduced in intensity. The sudden reduction of the intensity of the sun at sunset clearly shows the energy loss due to refraction. The additional energy loss due to scattering after the sun has set is also clear.

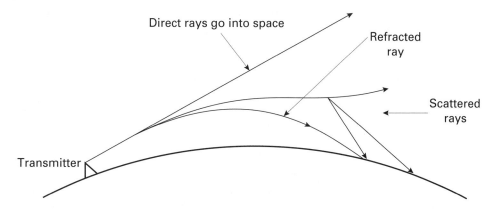

Figure 4–1.   Example of radio-wave propagation showing refraction and scattering.

Radio waves that are propagated through the atmosphere are called *sky waves.* At frequencies where both ground waves and sky waves can exist, both waves can be received by a receiver and the two waves will interfere. This phenomenon, called *fading,* affects the performance of several airborne navigation and communications systems. The frequency range where fading can occur depends on the time of day and the condition of the ionosphere.

Radio Noise

A universal enemy of radio communication is noise. Sources of radio noise are many, including natural and human-made. Much static is generated in the aircraft itself. In reciprocating engines, the main source of static is the aircraft's own ignition system. These problems have been known for a very long time, and shielded ignition systems have been used since radio equipment was first installed in aircraft. Other sources of static on board an aircraft must also be isolated in a similar fashion.

Another noise source from the aircraft is static charges that build up on the airframe. When the aircraft moves rapidly through the atmosphere, it accumulates a static charge. The charge will continue to build until the charge is such that the aircraft can suddenly discharge back to the atmosphere. The most common example of this phenomenon occurs in thunderstorms. The moving air masses build up charges in clouds that eventually discharge either to other clouds, the ground, or, in many cases, hapless aircraft. Lightning strikes to aircraft seldom cause human injury, but they can cause pitting of the aircraft and damage to electronic equipment. Discharging of an aircraft generates significant radio static, and it is necessary to provide an easy path for the aircraft to discharge.

Static discharges are caused when the electric field generated by the accumulated charge causes ionization of the air. Ionized air is a conductor of electrical current. The charge, having found a current path, quickly flows into the ionized air, which causes more ionization and creates a jagged, colorful trail of ionized air that drains off the charge of the aircraft. This is exactly the same process that occurs when a cloud formation builds a charge and then discharges either to the ground or another cloud, which is, of course, lightning.

When the aircraft discharges, a thunderous crash of static is heard in many aircraft radio receivers. The strength of the static is a function of the amount of charge that was dissipated. To keep the static level down, this charge level must be minimized.

The discharge will not occur unless the electric field is sufficient to start the ionizing process. Electric fields are strongest at the sharpest points of the aircraft. To keep the charge buildup to a minimum, very sharp points are deliberately introduced so that ionization occurs easily and very little charge can accumulate on the aircraft. These sharp points are provided by bundles of thin wire strands called *static dischargers.* They are placed throughout the aircraft, particularly at the areas of greatest airflow, such as the trailing edges of flying surfaces.

There are natural sources of radio noise and there is nothing engineers can do to reduce the level of this natural noise. A radio system must be capable of operating within the natural noise environment. Most natural noise takes place in the lower portion of the radio spectrum, that is, below about 3.5 MHz. This noise is generated from the normal electrical occurrences in the atmosphere such as thunderstorms and aurora. Electrical noise from thunderstorms is called *precipitation static,* or P-static.

Radio frequencies are chosen for navigation and communication systems based on their propagation characteristics and natural noise levels. Low frequencies, as an example, are used for long-range navigation because of predictable ground-wave propagation over long distances. Precipitation static is overcome by using sufficient transmitter power.

Once the electromagnetic energy is created by the antenna, the transmitter energy propagates away from the transmitter antenna at 3×10^8 m/s. As the energy travels away from the transmitter, it spreads out over increasingly wider areas, and thus the transmitted signal becomes weaker. If the energy is propagating in outer space where there are no losses, the energy is still there but spread out over a larger area. On Earth, if the propagation is straight line, or what is called *free space,* there is very little loss due to the air, and signal reduction is due to the increased area encompassed by the propagating energy. The energy density decreases as $1/R^2$, where R is the distance from the transmitter to the receiver.

The signal loss from transmitter to receiver due to free-space propagation is a function of distance only. Free space means that the loss of signal is from the spreading of the energy and not from such losses as generated by refraction or scattering. Equation (4-3) gives the path loss from the input of a transmitting antenna to the output of the receiver antenna.

$$\text{Loss} = \frac{\lambda^2 G_r G_t}{(4\pi)^2 R^2} \tag{4-3}$$

G_r is the gain of the receiver antenna, G_t is the gain of the transmitter antenna, R is the distance between the transmitter and receiver in meters, and λ is the wavelength of the transmission, also in meters. Usually, the gain of an antenna is given as a decibel figure relative to an isotropic radiator. An *isotropic radiator* can radiate equally well in all directions. Because the isotropic radiator is more of a theoretical antenna and nearly impossible to create as a practical example, antenna gains often are expressed relative to a dipole, which is easily fabricated. This permits antenna gains to be easily measured relative to the dipole reference.

Example

What is the loss between transmitter and receiver for two aircraft each using the same vertical antenna with a gain of 2.5 decibels (dB) over isotropic? The transmitting frequency is 122.8 MHz and the distance between the aircraft is 20 nautical miles (nmi).

Solution. The antenna gains must be converted from the decibel figure to a pure ratio

$$G_r = G_t = 10^{0.25} = 1.78$$

Next, the nautical mile separation must be converted to meters (m).

$$R = 1853 \text{ m/nmi} \times 20 \text{ nmi} = 37{,}064 \text{ m}$$

The wavelength is

$$\lambda = \frac{3 \times 10^8}{122.8 \times 10^6} = 2.44 \text{ m}$$

Substituting these values in Eq. (4-3) results in

101 dB of loss

The loss calculated is a dimensionless number, the ratio of the power density at a distant point compared to the power density at the transmitter. This represents an excellent use of the decibel because the loss can be very great.

It should be understood that the "loss" is not a loss of energy but a reduction in the available energy from a receiving antenna because of the spreading out of the transmitter's energy.

Example

Let's take the previous example a step further. A receiver can be easily built with a −100-dBm sensitivity. What minimum transmitter power would be required to achieve a distance of 20 nmi with this receiver?

Solution. A receiver sensitivity of −100 dBm when combined with a loss of 101 dB would require a transmitter power of +1 dBm, which is $10^{0.1} \times 1$ mW = 1.26 mW (milliwatts)

It would appear from the preceding example that minute amounts of transmitter power are all that are required for successful VHF communications. However, typical aircraft communications transmitters employ powers from 5 to 10 W. Five watts is 36 dB greater than the power required to establish communications according to the calculations of the preceding example. There are a number of reasons for the use of higher power. In the example we calculated the loss owing to the spreading of power due to propagation and considered no other losses. There are losses in the coaxial cables connecting the transmitter and receiver to their respective antennas, the antenna installation may be less than ideal and incur losses, there may be some shadowing effects by parts of the aircraft, there may be shadowing by structures and geographical features, and so on. Also, problems due to interference from noise sources and other radio users may be overcome by a stronger received radio signal. Finally, if communications is to take place at a distance greater than the geometrical line of sight where diffraction and scattering are involved, these two propagation effects, particularly scattering, involve considerable loss of signal. When all these effects are considered, the excess 36 dB of transmitted signal becomes necessary.

Photo 4–1   The Narco Com 811 TSO is a basic TSOed communications transceiver. Photo courtesy of Narco Avionics, Inc.

RADIO COMMUNICATIONS CONCEPTS

We have investigated the concepts of electromagnetic energy, a carrier, and applying the required information to the carrier. But before we can discuss the requirements of a radio communications system hardware, it would help to understand the need for communications. A communications system conveys information from one entity to another. In the case of aircraft, the communications is from ground to air and the reverse. What is the nature of our "information." Obviously it includes speech, but many other forms of information are transmitted, such as digital data, navigation information, and collision warnings.

Several forms of communications are in regular use in the aircraft industry: simplex, half-duplex, and full duplex. We are all familiar with these three types of communications techniques, but have probably never heard them called by their proper names.

Simplex is one-way communication such as in radio broadcasting. Many of us talk back to the radio or television but our responses are not heard. In the aviation industry, there are broadcast-style simplex communications such as weather broadcasts and information broadcasts for air terminals.

Half-duplex is two-way communications whereby all the participants take turns transmitting, but only one at a time. This is the typical two-way radio and represents most air-to-ground communications carried out today. For half-duplex communications to be successful, radio procedures must be adhered to so that only one party attempts to transmit at one time.

With full duplex, any user may transmit without interfering with any other user. The telephone is an example of this form of communication.

The three basic elements of a communications system are the transmitter, a communications medium, and a receiver. The transmitter takes the information to be communicated, prepares it to operate in the communications medium, and places the signal in the communications medium. The receiver plus its antenna retrieves the energy from the communications medium and restores the information to the same form as the original.

An example of this is speech, which enters the microphone of a radio transmitter as sound waves, is transmitted to the receiver through space, and emanates from the receiver's speaker as sound waves.

There is no argument that electromagnetic waves, that is, radio waves, are the communications medium of choice for airborne communications. Having established this, the roles of the transmitter and receiver are clear.

Most radio communications for airborne purposes is by amplitude modulation, and the majority of the communications takes place in the VHF spectrum from 118.0 to 135.975 MHz as shown in Table 4-1. Some communications, particularly for very long distances, use the high-frequency (HF) portions of the radio spectrum. Also, satellite communications is increasing. Satellites use a large variety of modulation methods. The common VHF transmitter must generate the desired carrier frequency, apply the voice information as amplitude modulation, and feed this energy to an antenna. The antenna converts electrical energy, the type of energy that travels through wires and coaxial cables, to electromagnetic energy, the type of energy that radiates through space.

TABLE 4-1 SPECIFIC FREQUENCY ASSIGNMENTS
FOR VOICE COMMUNICATIONS

Voice Communications Frequencies	
118.00–121.40	Air traffic control
121.5	Emergency
121.6–121.9	Airport ground control
121.95	Flight schools
121.975	Private aircraft advisory
122.0–122.675	Flight service station
122.7	Unicom
122.725	Unicom for private airports
122.75	Air to air
122.8	Unicom uncontrolled airports
122.85	Multicom
122.9	Multicom
122.925	Multicom
122.95	Unicom controlled airports
122.975	Unicom
123.0	Unicom uncontrolled airports
123.05	Unicom heliports
123.075	Unicom heliports
123.1	Search and rescue
123.15–123.575	Flight test
123.3	Flight school
123.5	Flight school
123.6–123.65	FSS or air traffic control
123.675–128.8	Air traffic control
128.825–132.0	En route
132.05–135.975	Air traffic control

The Radio Transmitter

Figure 4-2 shows a block diagram of an AM radio transmitter for speech communications. A source of carrier frequency is provided, and this signal is modulated with the voice input. The power of the signal is amplified to a level that will span the necessary distance, and the output is fed to the antenna.

Lest this text leave the student with the impression that radio transmitter design is as simple as the block diagram would leave us to believe, it must be stated that some of these simple tasks are not as easy as they appear.

Frequency Synthesizers

One such task is the carrier generator. It is the function of this block to provide the carrier frequency for the transmitter. The VHF communications band has 720 frequencies that must be generated. The accuracy of the carrier frequency must be 0.001% to prevent radio trans-

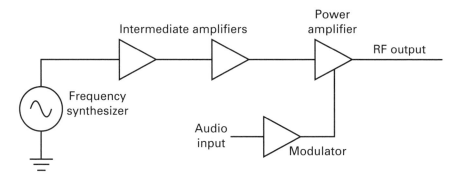

Figure 4–2. Block diagram of an AM radio transmitter.

mitters from interfering with each other. The device that performs this task is called a *frequency synthesizer* and tends to be a complex device. Frequency synthesizers are used in a number of aircraft electronics systems and will be encountered a number of times in this text.

One important electrical circuit that is used to generate a sine function for use as a carrier is an *oscillator.* A simple generator of a sine function was discussed in Chapter 2, the ac alternator. This is a mechanical device that generates a 400-Hz sine wave from the rotation of a shaft. In the very early days of radio before there were effective electronic amplifying devices, radio transmitters used hugh alternators to generate the required radio-frequency carriers. In those days "radio frequency" was extremely low compared to the frequencies used today, and even at these low frequencies, the alternators turned at very high rpms.

An oscillator is an electronic circuit that generates an alternating current with only electronic means. An oscillator circuit for providing a stable radio-frequency output uses a resonant circuit to set the frequency and an amplifier to replenish the energy removed from the circuit and keep the resonant circuit oscillating.

Resonant circuits may be constructed in a number of ways. In Chapter 2 we discussed resonance of inductors and capacitors connected together. Many oscillator circuits use the resonance of inductance, *L,* and capacitors, *C,* to set the frequency, and many oscillators are *LC* oscillators. However, there are a large number of applications for which *L*s and *C*s do not have the stability necessary to provide an accurate and stable output frequency. When extreme frequency precision and stability are required, a resonant circuit other than inductors and capacitors must be used. One suitable resonant device is the quartz crystal.

Quartz is one of a number of materials that exhibit a characteristic called the *piezoelectric effect.* Materials that exhibit this effect generate electricity when they are mechanically deformed. Also, they can be deformed if an electric field is applied to the device. If a disk of quartz is plated with a conducting material, applying an electric potential will cause the quartz to deform. The quartz disk has mechanical resonance, and if an ac electric field is applied at exactly the resonant frequency of the quartz disk, the quartz will oscillate with maximum amplitude. Although the oscillations are mechanical, the energy stored in these oscillations comes from an electrical source because of the conducting material on the quartz, and the crystal will appear as if it is an electrical resonant circuit.

The stability of the frequency of oscillation of the quartz disk is significantly greater than that of the best inductors and capacitors available. The best *LC* oscillator is able to provide a frequency stability over temperature and time of about 0.1% or one part in a thousand. Very inexpensive quartz crystals can provide stabilities 10 times better than the best *LC* oscillator, and crystal oscillators that are compensated for temperature effects can provide stabilities of 0.00001% or 0.1 part per million (ppm).

Practically every airborne transmitter or receiver requires the stability of a crystal oscillator, and crystal oscillators are very common. The chief disadvantage of the crystal oscillator is that there must be one crystal for each generated frequency. In the case of a communications transmitter with 720 operating frequencies, it is impractical to use one crystal per frequency. There are methods of reducing the number of crystals to 50 or 60 by using the crystal frequencies in combinations, but this is still an impractical number of crystals. In spite of the large number of crystals, some older airborne units used this technique, but these crystal multiplexing schemes are no longer used.

Because several airborne systems require a large number of accurate frequencies, systems of frequency generation called frequency synthesizers were developed that use one quartz crystal but provide an unlimited number of frequencies. The most common of the frequency synthesizer types is the *phase-locked-loop synthesizer.* The phase-locked-loop synthesizer uses an *LC* oscillator that is capable of providing a broad range of frequencies but would not have the necessary frequency stability for most aviation applications. The phased-locked loop gives the VCO the necessary stability by comparing the VCO's frequency to a stable crystal oscillator and making corrections as necessary.

Figure 4-3 shows the block diagram of a phase-locked-loop synthesizer. A special *LC* oscillator called a *voltage-controlled oscillator* (VCO) is used as the source of the RF signal. The voltage-controlled oscillator is a conventional *LC* oscillator in which a voltage-variable capacitor, which is a form of diode called a *varactor,* is used to tune the oscillator.

The output frequency is divided by an integer using digital circuits called a *programmable divider.* The division is by an integer that can be changed easily or programmed. If the division is *N,* then the output of the programmable divider is

$$F_{out} = \frac{F_{in}}{N} \qquad (4\text{-}4)$$

where F_{out} is the output frequency from the programmable divider and F_{in} is the input frequency.

Let us introduce another circuit called the *phase/frequency comparator.* This device has two inputs, F_a and F_b, and provides an output that indicates whether input F_a or F_b is higher in frequency. When F_a and F_b are exactly the same frequency, the output will now indicate whether F_a or F_b is leading or lagging in phase. There are a number of variants of this circuit but for the sake of clarity, assume that the output is positive when F_a is greater than F_b and remains positive when F_a leads F_b in phase. The output is negative when the opposite situations are true: F_b greater than F_a and F_b leading F_a in phase. The output is zero when F_a and F_b are exactly the same frequency and with zero degrees of phase shift.

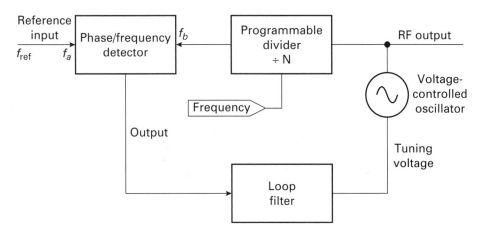

Figure 4–3. Block diagram of a phase-locked-loop synthesizer.

Let us also provide a very stable source of reference frequency called F_{ref}, which is derived from a crystal oscillator.

These elements are connected as shown in the block diagram of Fig. 4-3. The output of the phase/frequency detector is fed back to the VCO through a loop filter that has a very high dc gain. For the output of the loop filter to be finite, the input to the loop filter must be near zero. This requires that F_a and F_b be the same frequency and very nearly in phase. If this is to be true, then

$$F_a = F_b = F_{\text{ref}} = \frac{F_{\text{osc}}}{N} \tag{4-5}$$

or

$$F_{\text{osc}} = N F_{\text{ref}} \tag{4-6}$$

Therefore, the oscillator frequency may be set to any multiple of F_{ref} by changing the value of N.

Example

What value of F_{ref} and range of N would be required to provide the carrier frequencies for a VHF communications transmitter?

Solution. The VHF communications frequencies span from 118.00 to 135.975 MHz in 25-kHz steps. Because the frequency difference between channels is 25 kHz, this is the value of F_{ref}. For the lowest frequency, $N = F_{\text{osc}}/F_{\text{ref}} = 118.000$ MHz/0.025 MHz $= 4720$. For the highest frequency, $N = 135.975$ MHz/0.025 MHz $= 5519$. Therefore, N must assume values of 4720, 4721, 4722, . . . , 5519.

The output of the synthesizer is amplified and modulated before feeding the antenna. Just how much must the signal be amplified? What power level is required of an airborne

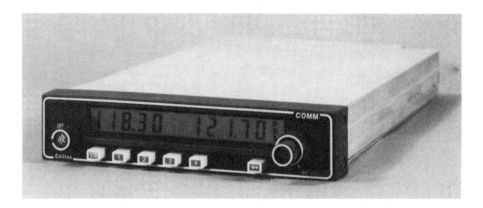

Photo 4–2   The Collins VHS 253 allows up to six frequencies to be stored and used.
Photo courtesy of Rockwell International Corp., Collins Divisions.

communications transmitter? This depends on a number of factors, but typical transmitter output powers are 5 to 10 W, and a typical synthesizer output is on the order of 10 mW, which requires a gain of 30 dB for the 10-W output.

COMMUNICATIONS RECEIVERS

The superheterodyne receiver, invented by Edwin Armstrong, was so superior to any previous receiver design, including some of those patented by Armstrong earlier, that it instantly became the only receiver topology worth using. This remains the case even today. Only very specialized receivers have designs that deviate from the superheterodyne topology. This chapter will discuss the superheterodyne receiver in detail because of its importance for all forms of radio-based aircraft systems.

Filters for Communications

Like frequency generation, filters are a very important part of communications electronics, particularly in a receiver. It would be prudent to review filters in this section.

There are four basic types of filter: low pass, high pass, band pass and band stop. Each filter is characterized by a transfer function that is a plot of the signal attenuation of the filter as a function of frequency. As an example, the ideal low-pass filter transfer function is shown in Fig. 4-4(a). This filter will pass all signals without attenuation below a particular frequency called the *cutoff frequency, F_c.* All signals above this frequency are completely attenuated. The area where signals are passed without attenuation is called the *passband* of the filter, while the area where signals are attenuated is called the *stopband*.

The high-pass filter is the mirror image of the low-pass filter and is shown in Fig. 4-4(b); it passes only those frequencies above a cutoff frequency, F_c. Signals below F_c are completely eliminated.

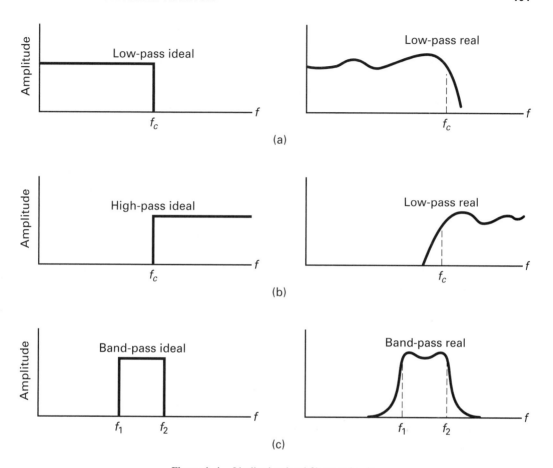

Figure 4–4.   Idealized and real filter passbands.

A band-pass filter is shown in Fig. 4-4(c); it passes all frequencies between two limits, F_1 and F_2. Outside the passband, all signals are eliminated.

A band-stop filter is the opposite of the band-pass filter. All signals between F_1 and F_2 are stopped, while all signals outside the stopband are passed without attenuation.

The transfer functions shown in Fig. 4-4 are ideal and real filters and do not exactly match these transfer functions. Just how a filter deviates from the ideal must be characterized in order to fully describe a real filter.

Figure 4-4 shows a transfer function of a practical low-pass filter. As can be seen from the figure, this filter does not look exactly like the ideal low-pass filter curve. First, the filter does not instantly transition from the passband to the stopband. Therefore, a criterion for setting the cutoff frequency F_c must be established. This criterion is at a point 3 dB from the attenuation of the passband. This brings forth two more defects of a real filter that must be resolved. First, the filter does not pass signals in the passband without loss; there is a finite loss in the passband called the *insertion loss*. Furthermore, the attenuation in the passband is not constant, but tends to have greater on lesser amounts of attenuation as a function of frequency. This variation in passband attenuation is called *passband ripple.* The insertion loss is the attenuation that lies midway between the point of least attenuation and the point of most attenuation. The cutoff frequency F_c is 3 dB down from the insertion loss.

Unlike the ideal filter passband, in which the filter transitions instantly from passband to stopband, practical filters have a finite slope. To quantify how rapidly the filter cuts off, the 3- and 60-dB attenuation points are specified. The closer the 3- and 60-dB points are, the more rapidly the filter makes the transition from passband to stopband.

Filters can be passive or active. Passive filters use inductors and capacitors, whereas active filters use resistors, capacitors, and amplifiers and are capable of making filters without inductors. The attraction to making filters without inductors is that inductors tend to be large, heavy, and expensive.

Active filters can be made only for lower frequencies, typically in the audio range. For radio frequencies, only passive filters using inductors and capacitors are used because of the limitations of active devices.

Very narrow band passive band-pass filters can be made with quartz crystals rather than inductors and capacitors. About the maximum bandwidth that can be achieved with quartz crystals is 0.1% of the center frequency. There are a number of requirements for narrow band-pass filters in avionics systems, including communications receivers that can take advantage of crystal filters.

Communications Receiver Characteristics

The communications receiver must remove the information that was applied to the carrier and provide that information to a loudspeaker or headphones, in the case of voice communications. To accomplish this, the receiver must provide the following.

> *Amplification:* Because of the tremendous loss of signal due to the spreading of energy during propagation, the received signal must be amplified by a significant amount.
>
> *Selectivity:* This is the ability of a receiver to amplify and process one signal while rejecting other signals.
>
> *Demodulation:* Information to be communicated is applied to a carrier for transmission; it is then necessary to remove this information by demodulation.

To appreciate the advantages of the superheterodyne receiver it would be helpful to study a common method of radio reception in use prior to the invention of the superheterodyne. Figure 4-5 shows a receiver topology called the *tuned radio frequency,* or TRF.

The signal from the antenna is fed to a band-pass filter, which provides the selectivity requirement of the receiver. After the filter, an amplifier is provided, after which follows a demodulator. After the demodulator, additional amplification is provided.

This receiver provides all the requirements of a radio receiver and appears simple from the block diagram, but it is essentially impossible to use this topology for VHF communications.

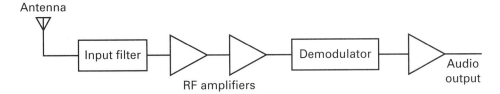

Antenna

RF amplifiers

Figure 4–5. TRF receiver, an early form of radio receiver.

The first difficulty encountered is the required bandwidth of the band-pass filter. A communications channel in the aircraft VHF communications band is 25 kHz wide, which is only 0.018% bandwidth at the highest operating frequency. Making a practical band-pass filter with this narrow bandwidth is virtually impossible. Also, the communications receiver must operate on 720 different frequencies, which requires that the band-pass filter be adjustable. This takes making the band-pass filter from virtually impossible to absolutely impossible.

To continue our discussion of the TRF receiver, let us assume that the TRF will not be used for VHF applications because of the problems with the band-pass filter, but at a lower frequency where the input band-pass filter can be realized. Let us now investigate the RF amplifier. Radio receivers are required to provide intelligible signals with input levels of around 1 microvolt (μV).

Example

For the purposes of investigation, assume that the receiver input resistance is 50 ohms (Ω), the standard resistance for practically every aircraft receiver, the input voltage to the receiver is 1 μV, and 10 W of audio power is required for the loudspeaker in a noisy aircraft cockpit. How much total receiver gain is required?

Solution. The input power is $(1 \times 10^{-6})^2/50 = 2 \times 10^{-14}$ W. The required gain is the output power divided by the input power or $10/(2 \times 10^{-14}) = 5 \times 10^{14}$ or 147 dB.

In the TRF, two frequencies are involved, the signal frequency, or the RF, and the demodulated frequency, which is audio frequency. The required 147 dB of gain must be divided between these two frequencies. Obtaining large amounts of gain at one frequency is a very undesirable practice because of the effects of noise and the possibility that some energy at the output of an amplifier may be fed back to the input and cause oscillations. Even if the required 147 dB of gain is divided equally into two 73.5-dB blocks, this amount of gain is excessive.

The superheterodyne receiver shown in Fig. 4-6 accomplishes the three requirements of a receiver very effectively. First, the input signal is filtered with a relatively broadband filter whose function will be discussed later. After the input band-pass filter, some gain is provided and followed by another band-pass filter. The input frequency is multiplied in a circuit called a *mixer* with an oscillator called the *local oscillator.*

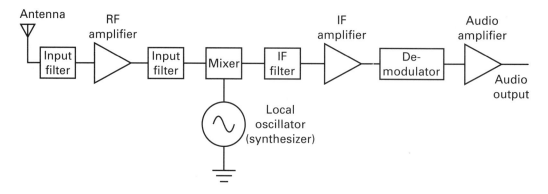

Figure 4–6. Block diagram of the superheterodyne receiver.

Whenever one frequency is multiplied by another, the result is two new frequencies, the sum and difference of the two original frequencies. The difference is called the intermediate frequency or the IF. The local oscillator frequency is typically above the input frequency by an amount equal to the IF. The IF filter, which has a bandwidth equal to the channel bandwidth, passes only the difference frequency and rejects the sum, which is higher than both the local oscillator and input frequencies. The IF signal is amplified, and after the amplification the signal is demodulated and amplified again.

The block diagram of the superheterodyne receiver after the mixer is the same as for the TRF: filtering, amplification, demodulation, and additional amplification.

What is so magical about first converting the input signal to an IF and treating the result the same as the TRF receiver? First, the IF is a constant frequency. The IF filter, which must be a narrowband filter, operates at only one frequency. This takes the absolutely impossible filter of the TRF design back to the realm of possible. Also, the IF is usually a lower frequency than the input frequency, which takes the IF filter further along from possible to practical.

Let us return to the fact that multiplying two frequencies results in both the sum and difference frequencies. This implies that the input frequency to the mixer is converted to two possible output frequencies, of which one is passed by the IF filter. It also means that two input frequencies may be converted to the same IF. This can be represented by

$$\text{(local oscillator frequency)} - \text{(input frequency 1)} = \text{IF}$$

$$\text{(input frequency 2)} - \text{(local oscillator frequency)} = \text{IF}$$

Adding the two equations results in

$$\text{(input frequency 2)} - \text{(input frequency 1)} = 2\text{ IF}$$

One of the two input frequencies is the *desired frequency* and the other is called the *image*. It does not matter which of the two frequencies is the desired frequency; the important fact is that there is a second frequency that can translate to the IF and cause difficulty. The filter at the input to the superheterodyne receiver eliminates image frequency signals.

The receiver IF is carefully chosen so that it is low enough that IF filters can be easily constructed, yet high enough that the image is easily removed.

Example

What is the local oscillator frequency range for a communications receiver with a 10.7-MHz IF when the local oscillator is placed above the input frequency? How far from the input frequency is the image frequency?

Solution.   The local oscillator covers from 118.00 MHz + 10.7 MHz = 128.7 MHz to 135.975 MHz + 10.7 MHz = 146.675 MHz. The image is above the input frequency by an amount equal to twice the IF, or 21.4 MHz, or a range from 139.4 to 157.37 MHz.

From the previous example, it can be seen that designing an input filter that can remove the image frequency 21.4 MHz from the desired frequency is not difficult.

The gain of the superheterodyne receiver can be achieved at any point within the receiver, but it is usually distributed throughout the receiver. Using our previous example of 147 dB of gain, the majority of the gain would be in the IF amplifier. However, unlike the TRF topology, the IF operates at only one frequency, which is typically a lower frequency than the RF. This allows relatively high gain to be employed without difficulty.

It is advantageous to distribute the gain throughout the receiver. A typical gain distribution would be

RF amplifier	20 dB
Mixer	15 dB
IF amplifier	60 dB
Audio amplifier	52 dB
Total	147 dB

One characteristic of a radio receiver is the need to operate with a wide variation of signal strengths. When an input signal is very strong, such as would occur when a received transmitter is very close, the audio from the speaker will be very loud, whereas more distant transmitters will produce a weak audio output. Since an aircraft receives transmissions from a number of aircraft and ground facilities, there will be large variations in the audio output. A form of automatically adjusting the gain of the receiver to accommodate the range of signal levels is provided for. This system is called *automatic gain control* or AGC.

Example

What is the difference in the signal strengths received by an aircraft from a tower 1000 feet (ft) distant and an aircraft 50 nmi distant at an altitude of 5000 ft above ground level (AGL). Assume the aircraft and tower transmit with a power output of 10 W and both use dipole-equivalent antennas.

Solution.   Because the aircraft is 50 nmi and 5000 ft AGL, it would be safe to assume that the path between the two aircraft is not over the geometrical horizon and that free-space loss is involved. Therefore, to find the ratio of the losses, Eq. (4-3) is used. The ratio of the losses results in a simple relationship.

$$\left(\frac{R_1}{R_2}\right)^2 = \left(\frac{3.04 \times 10^5}{10^3}\right)^2 = (304)^2 = 9.24 \times 10^4$$

This equation involves the ratio of the distances squared and thus the difference in signal strengths is 50 dB.

Fifty decibels is not a large difference in signal strengths for a radio receiver. A 50-dB difference in audio output from the receiver would be a rather unpleasant experience to a listener, but the AGC system would equalize the audio from the tower and aircraft. It is not uncommon for receivers to be required to operate with signals that differ by much more than 50 dB. If the airborne aircraft were over the geometrical horizon, there would be additional losses over the free-space loss.

The AGC system uses a variable-gain IF amplifier by which the gain is controlled from a feedback signal from the demodulator. Since the carrier of an AM signal is unaffected by the modulation, the carrier is a reliable indicator of the received signal strength. When an AM signal is detected, the dc component is proportional to the carrier and is used to control the gain of the receiver.

Because the IF amplifier is the part of the superheterodyne receiver that contains the most gain, it is also the best source of gain control for the AGC system. The variation of input signal strength can be as great as 100 dB. In the proposed gain distribution, the gain of the IF amplifier is only 60 dB, and if the IF amplifier were the sole source of gain control, the IF amplifier would have to range from a gain of 60 dB to a loss of 40 dB. Because of the large variation in gain, it would be desirable to obtain gain reduction from both the IF amplifier and the RF amplifier.

Squelch and SELCAL

A radio receiver that is not receiving a signal produces a noise output that is annoying to the operator. This noise consists of noise generated in the receiver, as well as weak unreadable signals on the channel that the receiver is tuned to. A *squelch circuit* is provided in the receiver to remove the audio output when no signal is being received.

Although there are a number of methods of implementing a squelch circuit, one common method is to use the AGC control voltage. The AGC control voltage is a function of the strength of the received signal. When the received signal is strong enough to cause AGC action, this is sensed, and the squelch circuit allows the audio to be applied to the speaker.

In some situations more than just a squelch system is needed to control the receiver. When at a terminal it is important that all aircraft hear all communications taking place. But when en route, communications take place that are not associated with aircraft maneuvers. These usually involve air transport flight information relative to the number of passengers on board, the need for wheel chairs, time of arrival, times of connecting flights, and so on. If every air transport for a particular airline was constantly bombarded with these communications, it would be distracting for the aircrew. Therefore, a system of selective calling, called SELCAL, is used in larger aircraft.

Photo 4–3 The airborne control panel for the Air Transport Selective Calling and Data System. This unit provides data transmission for airlines as well as selective calling. Photo courtesy of Coltech, Inc.

Selective calling systems use audio-frequency tones that are transmitted before each transmission. Each receiver decodes the tones, and if the tone sequence matches that assigned to the receiver, the audio output is enabled and the communications is heard.

Because the activity on a communications channel is not heard when the SELCAL is enabled, before transmitting, it is necessary to disable the SELCAL, thus allowing the receiver to be activated. This is usually accomplished by providing a hook switch on the microphone so that when the microphone is removed from its hanger the audio of the receiver is enabled.

Communications Transceiver

A VHF communications system can use a separate transmitter and receiver or may combine both functions into one unit. There are cost advantages to combining the transmitter and receiver circuits. These cost advantages are gained by sharing circuits between the transmit and receive functions. Not every circuit can be shared, but some expensive circuits may be shared. Figure 4-7 shows a VHF communications transceiver that shares transmit and receive circuits.

The most effective circuit sharer is the synthesizer. When receiving, the synthesizer provides the receiver local oscillator, and while transmitting, the synthesizer provides the transmitter carrier frequency. These are two different frequencies, and it is necessary that the synthesizer be able to rapidly switch from the receiver local oscillator frequency to the transmit carrier frequency when the microphone push-to-talk switch is pressed. This requirement complicates the synthesizer design a bit, but it is worth the effort.

Another way that circuits may be shared is when the audio amplifier serves as the modulator for the transmitter. Most VHF communications transceivers provide an output power between 5 and 10 W, which requires a modulator capable of providing between 2.5 and 5 W of audio power. This power level is commensurate with the power required for a cockpit speaker, and the receiver's audio amplifier can serve as the transmitter's modulator.

In other areas, such as power supplies, sharing can take place, and, of course, since both transmitter and receiver reside in one housing, all the mechanics, including plugs, jacks, and mounting hardware, are shared. This is a significant saving in cost and weight.

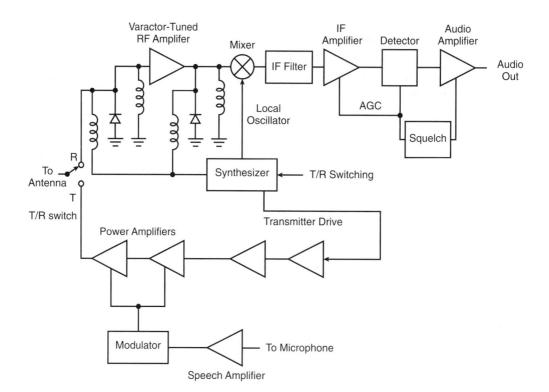

Figure 4–7.   Block diagram of a communications transceiver.

High-frequency Radio Communications

VHF communications can cover a range of 200 nmi, maximum, at cruising altitude. There are many requirements for communications beyond this distance, such as in transoceanic flights and over sparsely inhabited areas of the world such as northern Canada. For these communications requirements, high-frequency radio is used. As explained previously in this chapter, high-frequency radio waves can be propagated great distances because the radio energy is refracted and reflected from the ionosphere.

The range of frequencies used for high-frequency communications spans from 3 to about 20 MHz. The mode of transmission is single sideband, or SSB.

When a carrier is amplitude modulated, two sidebands are generated, an upper and a lower, as shown in Fig. 4-8. Even though the carrier is modulated, from a spectrum standpoint it remains unchanged from its form before modulation. The bandwidth of the AM signal is two times the modulating frequency because the signal contains two sidebands, each removed from the carrier frequency by an amount equal to the modulating frequency. For speech modulation, the maximum modulating frequency is about 3 kHz and thus the bandwidth is 6 kHz. Theoretically, the amount of spectrum required to transmit a modu-

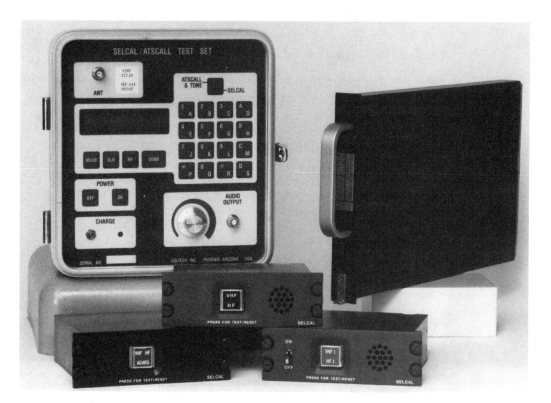

Photo 4–4 A collection of SELCAL equipment showing a number of decoders and a test set. Photo courtesy of Coltech, Inc.

Unmodulated carrier

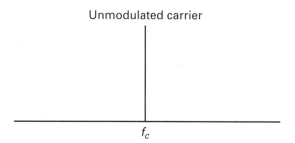

f_c

Amplitude modulation

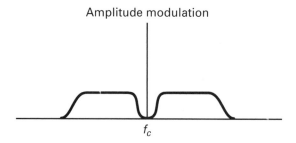

f_c

Single sideband, lower sideband

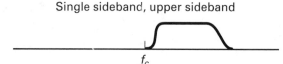

f_c

Single sideband, upper sideband

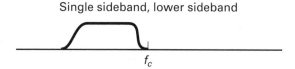

f_c

Figure 4–8. Spectrum of an amplitude-modulated carrier showing single sideband.

lating waveform is that which is equal to the maximum bandwidth of the modulating signal. Speech spans from 300 Hz to 3 kHz for a total bandwidth of 2.7 kHz. The 6 kHz required for the amplitude-modulated signal is more than twice the theoretical maximum because there are two sidebands. The same information is contained in each sideband, and no information is contained in the carrier. The required spectrum could be reduced to the theoretical minimum if the carrier and one sideband were removed.

VHF communications frequencies may be reused at distances of about 200 nmi. This is because beyond 200 nmi aircraft are out of range. In the high-frequency spectrum, the range is much greater, which requires more users to share communications channels. Because of this, it is necessary that the minimum bandwidth be used for communications to allow for the maximum number of channels; therefore, only one sideband is transmitted. This type of transmission is called *single-sideband suppressed carrier*, or SSB for single sideband.

Either sideband works equally well; the upper sideband is designated USB and the lower sideband is LSB. LSB is not compatible with USB, and when operating SSB the sideband must be specified. As an example, if an operating frequency is specified as 7.773 MHz, it must include the sideband designator. Therefore, the operating channel designation would be 7.773 MHz, USB.

SSB transmission and reception require accurate tuning of the receiver for an intelligible reception. Typically, the HF transceiver provides frequency-setting capability to within 100 Hz. The frequency coverage of the typical HF transceiver is from 3 MHz to 20 MHz, and in some cases 30 MHz, which represents up to 270,000 frequencies.

ONBOARD COMMUNICATIONS

It is clear that radio is the choice of communications medium for air to ground communications, but there are requirements for considerable onboard communications. These requirements are for such applications as voice communications among flight crew members and between flight crew members and passengers. This communication is provided by a system called an interphone and usually works in conjunction with an audio panel, which not only provides the interphone function but switches speakers and headphones between the various communications and navigation systems aboard the aircraft.

Another very important communications requirement on board an aircraft is among the various systems in the aircraft. This is data communication and, of course, is strictly electronic.

Audio Panel

There are a number of audio sources in the aircraft that the flight crew should monitor. Obviously, the communications transceivers must be heard to be of any value. Typically, there are two and even three transceivers on larger aircraft, and there should be methods of selecting transceivers that are to be heard and transceivers that are to be used for transmitting.

A number of the navigation equipment also have audio outputs for station identifiers. These navigation systems are ADF, VOR, localizer, marker beacon, DME, and Microwave Landing System (MLS). In the surveillance equipment, only the Traffic and Collision Avoidance System (TCAS) systems have an audio output.

Realizing that some aircraft have two of some navigation systems, there must be a method of selecting which navigation system's audio will be heard in the speaker or the headphones. This function is achieved by the audio control panel.

Figure 4-9 shows a typical audio control panel. The actual method of implementing the audio control panel can vary from what is shown. There is no standardized system of implementing the audio switching in general aviation audio control panels. However, some common features are evident in most panels. As a matter of interest, it is this kind of unnecessary confusion that ARINC specifications prevent. (The function of ARINC was discussed in Chapter 1.) An audio control panel manufactured by one manufacturer to an ARINC specification operates exactly the same as those from all other manufacturers who use the same specification. This prevents a dangerous situation because of unfamiliarity with the equipment.

Figure 4-9 shows two rows of selector switches; one row selects sources for the speaker, while another row selects sources for the headphones. Some audio sources are not switched, such as the voice warning signals from the collision avoidance system and the marker beacon used for instrument landings.

Often, the marker beacon has provisions for temporarily removing the audio so that the beeping of the marker beacon as an aircraft makes an instrument approach does not interfere with other necessary audio sources, such as the communications radio. This function is called marker mute and employs a timer that will restore the audio after 10 s or so such that the audio is restored before the next marker is passed. In many audio panels for smaller aircraft, the marker beacon receiver or at least the marker beacon indicators are mounted in the audio control panel.

The audio control panel must provide switching for the communications transceiver transmit function. Most aircraft contain at least two communications transceivers, and only one can be controlled from the microphone. This switching function must be handled by the audio control panel.

A block diagram of a control panel is shown in Fig. 4-10. Essentially, the control panel contains switching and amplification.

Typically, the audio control panel contains the speaker power amplifier. Many communications transceivers have a speaker amplifier so that they may be used either without an audio control panel or with a simple audio control panel that does not have a speaker power amplifier. As explained earlier, many transceivers use the speaker amplifier as a modulator for the transmitter, so the amplifier is needed for the transmitter. Navigation receivers, on the other hand, gain no benefit from having a power amplifier, and most navigation receivers have no speaker amplifiers. When communications transceivers and navigation receivers are contained in one package, the speaker amplifier for the navigation receiver is provided by the communications transceiver. These NAV/COM units are typically found on smaller aircraft.

In larger aircraft, it is not uncommon to find that none of the communications equipment or navigation equipment is equipped with speaker amplifiers, but relies on a speaker amplifier in the audio control panel. The same is true of a headphone amplifier, and the audio panel contains a headphone amplifier as well.

Figure 4–9. Example of an audio control panel.

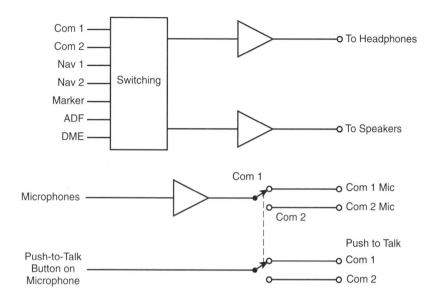

Figure 4–10. Block diagram of an audio control panel.

Switching is accomplished in the audio control panel either with a conventional mechanical switch or by various electronic means. Switching must be done with care to prevent various undesirable conditions. The first is *crosstalk*. If, as an example, a NAV receiver is selected for the cabin speaker and a COM radio is selected for the headphones, no signal from the NAV radio should be heard in the headphones. This should also be true for any unselected source.

Another characteristic is *isolation*. This is when one source is selected for, say, the speaker, and then the same source is selected for the headphones; there should be no change in audio level when the source is added to the headphones.

The microphone must be switched to the desired communications transceiver. When switching the microphone, both the audio from the microphone and the push-to-talk switch must be routed to the selected transceiver.

Also, in many larger aircraft, there are a number of microphones such as pilot and copilot and hand microphones and boom or (oxygen) mask microphones.

The audio panel is a very critical piece of equipment aboard the aircraft. Since all the audio sources pass through the audio panel, a failure in the audio panel can cause a serious problem. Usually, the loss of all NAV audio, which would be a loss of the identifiers for the NAV aids, is a situation that can be tolerated, but the loss of all radio communication is critical. Therefore, many audio control panels have an emergency mode.

The emergency mode permits bypassing the audio panel and allows one transceiver to be operated. In many cases, since the speaker amplifier is in the audio panel, there will be no cabin speaker and headphones must be used. To reduce the likelihood of an audio panel failure, many larger aircraft provide two audio panels, one for the pilot and a second for the first officer.

Communications Microphones

Good readable communications is not likely if a good quality microphone is not used. A distorted signal from the microphone is fed to the transmitter, where an equally distorted signal is transmitted and received.

A number of different types of microphones are in use in aircraft communications today. One microphone type that is falling out of favor, but that was the mainstay for a number of years, is the *carbon microphone.* This type of microphone has a diaphragm that is connected to a capsule of carbon granules, as shown in Fig. 4-11. The sound pressure from speaking into the microphone deflects the diaphragm and compresses the carbon granules in the capsule. This decreases the resistance of the carbon granules in synchronism with the sound pressure.

The carbon microphone is connected in series with a resistor as shown in Fig. 4-12, which creates a voltage divider. Because one resistor in the voltage divider is a variable resistor, the output voltage is a function of the sound pressure, and hence the audio-frequency waveform of the speech appears at the resistor.

The carbon microphone does not produce high-quality speech output, but it has been used for many years because of the extreme ruggedness of the microphone. For speech, the microphone is acceptable, and it was the microphone originally developed for and used in telephones for over 100 years.

An alternative form of microphone, which originally was not as rugged as the carbon microphone, is the *dynamic microphone.* The construction of this microphone has steadily improved to the point where its ruggedness is on a par with the carbon microphone. The distortion level of this microphone is much lower than for the carbon microphone, and this type of microphone has rapidly replaced the carbon microphone.

Figure 4-13 shows the construction of the dynamic microphone. Sound pressure moving the diaphragm causes movement of the coil in the magnetic field, which produces a voltage in the coil.

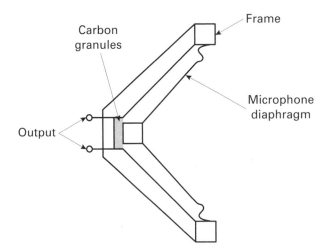

Figure 4–11. Cutaway view of a carbon microphone.

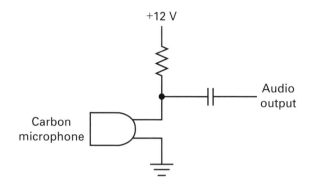

Figure 4–12. Connections for a carbon microphone.

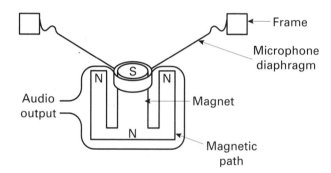

Figure 4–13. Construction of a dynamic microphone.

Many dynamic microphones are made to replace carbon microphones, and one of the significant differences between the carbon and dynamic microphones is the level of audio voltage output. The carbon microphone was capable of very high levels of audio output voltage. In the early radio transmitters, this represented a considerable saving in weight and power consumption because fewer audio amplifier stages were required. In modern equipment in which one integrated circuit provides more than enough gain for a dynamic microphone, carbon microphones are seldom used. However, because the carbon microphone was used for a long period of time, most aircraft are equipped to use the carbon microphone, and newer radio transceivers are made to be compatible with the existing microphones. To bring the audio level of the dynamic microphone up to that provided by a carbon microphone, the dynamic microphone is provided with an amplifier within the microphone case. The power for the amplifier is provided by the normal battery supply and series resistor used in a carbon microphone. Therefore, the dynamic microphone can substitute for the carbon microphone.

The use of the amplified microphone to replace the carbon microphone is so widespread that amplified microphones are now the procedure accepted for aircraft.

A more recent microphone makes use of a type of material called an *electret*. This classification describes materials that are insulators and have trapped charges. If the material is used for the dielectric of a capacitor, it is like creating a charged capacitor without applying a voltage to charge the capacitor. Electret materials are sometimes called *ferro-electric* materials, because the residual charge in an electret is the electric-field equivalent of a residual magnetic field in a ferromagnetic material.

The ferroelectric material is used as a dielectric for a capacitor in which the capacitor plates are the frame of the microphone and a diaphragm, as shown in Fig. 4-14. This results in a charged capacitor without the introduction of a charging voltage. In addition, the capacitance is a function of the distance between the diaphragm and the frame, which is a function of the sound pressure falling on the diaphragm.

The voltage across a capacitor is

$$V = \frac{Q}{C} \tag{4-7}$$

However, for a parallel-plate capacitor the capacitance is inversely proportional to the distance separating the capacitor plates, and Eq. (4-7) is amended to be

$$V = kQd \tag{4-8}$$

The constant, k, involves the equivalent dielectric constant of the electret material and the area of the plates. The output voltage of this capacitor is a function of the sound pressure changing the distance, d.

The electret microphone has a very high output impedance because in practical microphones the capacitance of the diaphragm is very low. This requires a very high impedance load for the microphone. To prevent noise pickup by the high-input impedance amplifiers, an FET preamplifier is provided within the microphone and provides a low output impedance. This requires that the electret microphone receive power. The voltage supplied to the electret microphone is for the preamplifier; the electret material provides a capacitor charge without application of a voltage.

Early problems with electret material limited the temperature range and shock survivability of this microphone. Recent advances have eliminated many of the earlier problems, and electret microphones are finding widespread use in the aviation industry.

A special microphone construction technique called the *noise-canceling microphone* is shown in Fig. 4-15. In this microphone, two ports are provided for sound entry. One entrance hole is provided in the front of the microphone, and a second hole is provided in the rear or side of the microphone. The front entry provides sound pressure to the front of the diaphragm, and the rear or side entry provides pressure to the rear of the diaphragm. Background noise in the aircraft enters from both the front and rear holes, and the same amount of sound pressure falls on both the front and rear of the diaphragm. Because the pressures are the same on both sides, the diaphragm does not move and thus the background

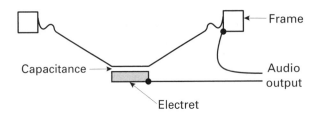

Figure 4–14.   Construction of an electret microphone.

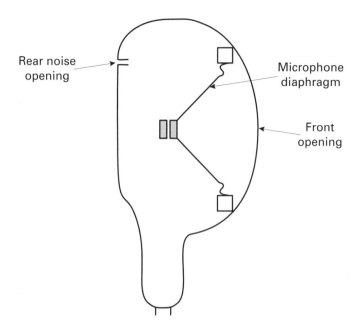

Figure 4–15.   Cutaway view of the noise-canceling microphone.

noise is suppressed. It is very important that the rear noise entry be sufficiently removed from the front entry, because if a significant amount of the desired speech energy enters both openings, it will be canceled as noise.

When speech enters from only the front opening, which requires that the speaker talk directly into the microphone, there is considerably more sound pressure from the front than from the rear opening, and the diaphragm will deflect and thus generate an output.

Noise canceling microphones can operate with any microphone technology and are effective at reducing background noise; they find widespread use in aircraft systems.

SPEAKERS AND EARPHONES

Cabin speakers and earphones use the same principle as the dynamic microphone, but in reverse. Other principles are used for speakers, such as the electrostatic speaker, which is the same as a capacitor microphone and is similar to the electret microphone, but this speaker design is impractical for aircraft use. Some earphones use the piezoelectric effect, but these devices are very fragile and not used in aircraft.

The cutaway view of a dynamic earphone or cabin speaker is exactly the same as for the dynamic microphone. Instead of sound pressure moving the diaphragm and causing current to be induced in the coil, current is passed through the coil, causing the diaphragm to move, which causes sound pressure. The dynamic earphone is similar in size to the dynamic microphone element, but the cabin speaker uses a diaphragm that is between 6 and 8 in.

Onboard Digital Communications

The modern trend in avionics is to interconnect much of the avionics equipment to create an avionics system. Essentially, a computer, the navigation computer, controls the avionics equipment, now called *sensors,* and uses the data from the sensors to solve the navigation problem. This results in the use of several navigation systems to arrive at enhanced navigational guidance. To create a large system such as this, a fair amount of communication must take place between the sensors and the navigation computer. Both analog and digital onboard communications have been present in aircraft for some time. However, the growth of digital navigation systems and the advantages of error detection and correction have resulted in, essentially, only digital communications aboard an aircraft.

The nature of digital communications is varied. Some of the requirements for digital communications involve what is already digital information. As an example, many communications transceivers, particularly in larger aircraft, are mounted in an avionics rack while the functioning of the transceiver is controlled from the flight deck. The setting of a frequency is a perfect example of a digital control function that needs to be communicated to the transceiver in the avionics rack. The volume control or the audio from the communications transceiver is an analog function. To transmit these functions, the analog signals must be digitized into numbers and transmitted digitally. Most of the digital communications that takes place in aircraft is of information that is inherently digital. Indeed, many of the more modern aircraft avionics systems designs are contemplating all-digital intercommunication.

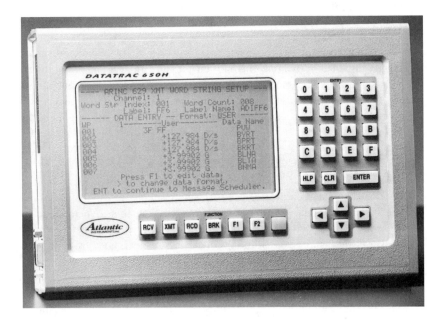

Photo 4–5   A test set for evaluating ARINC 629 data systems. Photo courtesy of Atlantic Instruments, Inc.

Because sensors may be produced by a number of manufacturers, some form of standard communications interface is required for compatibility. To ensure complete compatibility, a number of parameters of the communications system must be standardized.

First, the electrical characteristics must be specified. This includes the voltages for logic 1 and 0, the type of interconnecting cables, the minimum resistance that a receiver can have, and so on.

The data format must also be specified. This specifies how long a pulse shall be, just how much time must elapse between the transmission of data words, and so on.

Finally, the protocols must be specified. These include the length of words, the labels that identify data, parity bits, and so on.

One widely used onboard digital communications system is configured to the specification ARINC 429. As outlined in Chapter 1, ARINC is an organization that serves the airline industry. However, some ARINC specifications that are useful for general aviation (GA) have been adopted by the GA industry. ARINC 429 is a perfect example of an airline specification that has gained widespread acceptance in the general aviation industry.

ARINC 429 describes a digital communications system that operates at two speeds. The slow-speed ARINC 429 provides a nominal 12-kHz bit rate, whereas the high-speed ARINC 429 provides a 100-kHz bit rate. It can be seen from the speeds involved in ARINC 429 that it is used for the purpose of transmitting data that do not change at a rapid rate, such as digitized speech. ARINC 429 was specified long before fully digital systems were ever contemplated.

ARINC 429 operates with a type of code called a *return-to-zero* or RZ code, which is shown in Fig. 4-16. There are several significant advantages to this type of code. First, when no data are being transmitted, there is no signal present on the line, which saves energy. Second, each pulse has a positive and negative transition from zero volts to either a positive or negative voltage. Therefore, the first transition can be used to clock the state of the pulse voltage. This makes the RZ code a self-clocking code.

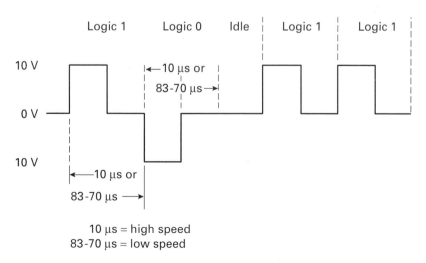

Figure 4–16.   An example of data transmission by ARINC 429.

The data words may have any word length, and the completion of a data word is signified by providing no pulses, positive or negative, for 4 bit times. A bit time refers to the nominal time required to send one data bit. The ARINC 429 data receiver recognizes this lack of activity and marks the end of the data word.

Each data word is identified with a label. This is a group of bits that indicates what data the word contains. As an example, magnetic heading would be a label. Also, frequencies would be a label. In the case of frequency, some frequencies require two data words because of the large number of significant digits involved. Therefore, the data label would specify the high- or low-order digits.

ARINC 429 is a one-way system in which only one transmitter is on any transmission line. An example is a DME transmitter/receiver mounted in the avionics rack that provides other equipment with data. Some of the data words from the DME would be distance, ground speed, time to station, identity, and the state of the system flag. The equipment that would be interested in receiving this information would be the DME display, an air data computer, an autopilot, and a navigation computer. The DME drives a shielded twisted-pair transmission line, and the desired receivers would tap the data from that transmission line. Not all the data are required by the receivers. As an example, the air data computer requires ground speed to perform some of its calculations, but the identity of the station or the actual DME distance or time to station are of no value. Therefore, the air data computer will decode only the ground speed and ignore other data words.

To reduce the possibility of receiving corrupted data, the ARINC 429 employs a parity bit. Parity is a rather ineffective error-detection scheme, and much improved error-detection methods are available. However, the ARINC 429 system is well protected from interference, and the system was invented before practical hardware was available for implementing the more sophisticated error-detection schemes.

The data are transmitted over a balanced differential transmission system using twisted pair. This type of transmission system helps to reject noise induced in the transmission line. The differential system operates with the voltage across the two lines as shown in Fig. 4-17. The balanced system subtracts the voltage of one side of the line from the other, which results in the difference voltage across the line. Using the voltage across the line makes the system immune to voltages generated in the line from external magnetic or electric fields.

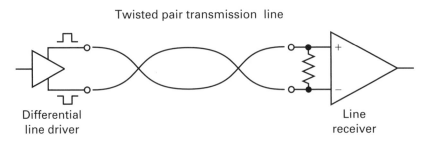

Figure 4–17.   Example of the ARINC 429 transmission system.

Let's go back to the transformer. Transformer action occurred because voltage was generated in a wire, which was the secondary of the transformer, from the magnetic field generated by the primary. If a wire is in a magnetic field, which would occur if a wire were near a motor, or routed along a wire carrying heavy current, or in the electromagnetic field of a radio transmitter, or anywhere in the aircraft when it takes a lightning strike, voltage will surely be introduced into that wire.

If the wires carrying the communications signal are subjected to the same magnetic or electric field, the voltage induced in the line will be the same for both lines. Since the signal is the difference of the individual line voltages, the same induced line voltage will appear at each line, resulting in no change in the line-to-line voltage.

This canceling effect is only valid if each side of the transmission line is subjected to the same fields. One way of assuring this is to twist the wires together so that they will not separate. This is the origin of the "twisted pair."

The ARINC 429 data transfer system is an older design, and although it continues to serve well in many applications, it has a number of shortcomings. Its two major problems are, first, it is very slow. Even the "high speed," 100-kHz version is slow by modern standards. Second, the one transmitter per line restriction severely limits the flexibility of the system.

An improved onboard data communications system is designated the ARINC 629. First, this data link operates at 1 MHz, still not fast by data communications standards, but sufficient for most applications. Second, the ARINC 629 system allows for multiple transmitters on a transmission line, which greatly improves flexibility.

Another interesting difference in the newer 629 specification is that it allows for the use of fiber-optic data transmission. This can save weight and can vastly improve the ability of the system to withstand induced noise.

FIBER OPTICS IN AIRCRAFT

Every effort has been taken in the design of digital communications systems used in aircraft to permit the system to be immune from the effects of radiated electromagnetic fields. No electrical system is ever fully immune from electromagnetic radiation, and even ARINC 429 and ARINC 629 can be affected. Some communications links carry such sensitive data that even temporary outages can cause serious difficulties. Such systems are autopilots and fly-by-wire systems in which the control of the aircraft is through electronic means, which require significant data communications.

Photo 4–6 The Terra TX 3200 communications transceiver is capable of storing and recalling ten frequencies. Photo coutesy of Terra Corporation.

There are a number of unavoidable sources of significant electromagnetic radiation aboard an aircraft. Several sources are the radio transmitters, such as the communications transceivers, DME, transponder, TCAS, radio altimeters, and weather radar. Another source that can produce enormous field intensity is a lightning strike.

Although a lightning strike is rare, it is common enough that most transport category aircraft are struck 10 or more times during their life. A lightning strike does very little damage other than startling the passengers and pitting the aircraft skin where the strike occurred. However, if an aircraft is controlled through an electronic data link that is severed during a lightning strike, the effects could be catastrophic.

These sources of high-voltage fields have been lumped into a category called high-intensity radiated fields (HIRF). A technique that will virtually eliminate the susceptibility of the communications system to HIRF is the use of fiber optics. The technique used in fiber-optic communications permits data to be transmitted by light waves through a very thin glass fiber. The light energy is modulated by turning the light source on and off with the data to be transmitted. The light energy is directed through the fiber much like water is constrained to flow in a pipe.

The mechanism that keeps the light energy constrained to the glass fiber is called total internal reflection. A characteristic of any transparent material is the index of refraction. The index of refraction is the ratio of the speed of light in a vacuum to the speed of light in the transparent material. When light energy passes from a material of one index of refraction to material of another index of refraction, some of the light energy passes through the boundary between the two indexes and some is reflected back. If the angle between the light energy and the boundary is less than what is called the *critical angle,* all the light energy is reflected and no energy passes through the boundary. When the light energy is within a thin glass fiber, none of the energy can escape the fiber, a phenomenon called *total internal reflection.* If a material is shaped into a very thin fiber so that the light energy is constrained to a very narrow passage, the angle between the surface of the fiber and the light rays is very small and less than the critical angle. This causes all the energy to be reflected and contained within the fiber.

The optical fiber is similar to a transmission line. Recall that the transmission line is constructed such that the electric and magnetic fields are totally contained so that no energy can escape. The transmission line is then used for transmitting signals with little loss and freedom from signals leaking into the transmission line. This energy containment works both ways; not only is energy constrained to travel along the transmission line, but external energy is precluded from entering the transmission line and corrupting the signals being transmitted along the transmission line.

The optical fiber is used in aircraft applications both for protection against electromagnetic interference and as a method of reducing the interconnecting cables, which results in weight reduction. A general technique for decreasing the weight created by a large number of interconnecting cables in an aircraft is called *multiplexing.* This technique places a number of signals on one wire, usually a transmission line. It is possible to place literally hundreds of signals on one transmission line and thus reduce the number of wires and connectors and achieve significant weight reduction.

A number of techniques are used to multiplex signals, but one characteristic is common to all methods; the more signals that are multiplexed, the greater the bandwidth of the composite signal. How is bandwidth related to the data rate of a transmission system? The faster that data are transmitted, that is, the more bits per second that are transmitted, the greater the bandwidth of the transmission. There is a direct relationship between the data rate and the bandwidth for any type of transmission. Thus, if the data rate is doubled, the required bandwidth to transmit the new data rate is also doubled.

A major factor of a transmission system is the number of bits that can be transmitted each second for each hertz of bandwidth. Even a simple data transmission system can achieve 1 bit per second for each hertz of bandwidth. More sophisticated systems can transmit 4, 8 or even more bits per second per hertz of bandwidth. There is a price to pay for this sophistication. Because the transmission system transmits a higher data rate per hertz of bandwidth, it is more susceptible to degradation by noise. Simpler systems can operate in the presence of greater noise energy, but at the expense of bandwidth.

With multiplexing, a number of users share a data transmission system. Multiplexing requires a data transmission system of greater capacity than for the data of only one user. As an example, if two users are to share a data transmission system and each user must pass 100,000 bits of data per second, the transmission system must be capable of passing 200,000 bits per second.

In the ARINC 429 system, for each transmission line there is only one transmitter, and thus the transmission line needs to be able to handle only the data of one transmitter. Although there are a number of receivers, their presence does not require an increase in the data-handling capability of the transmission system.

When there are a number of transmitters, which is a characteristic of ARINC 629, each transmitter must share the transmission medium. ARINC 629 uses a type of multiplexing called time-division multiplexing. Only one transmitter is on the transmission line at any one time. Although there may be a number of receivers, each receiver only uses the data that it needs. There are a number of methods of ensuring that only one transmitter is on the line at any one time. A specific time slot may be allocated to a particular transmitter and timing signals provided to synchronize the transmitters. Another method is to allow any transmitter desiring the line to use the transmission line when necessary. To use this type of unsynchronized sharing, the data transmission must include error-detection capability in case two transmitters try to transmit on the line at the same time. This situation is called a *collision,* and when a collision occurs, the data must be retransmitted. The transmitter must know that the data were not received, and the receiving end must acknowledge the receipt of a valid data transmission. This technique of error detection and acknowledgment is called *hand-shaking.*

Frequency-division multiplexing and code-division multiplexing are more sophisticated types of multiplexing. In addition, a number of enhancements to these basic multiplexing schemes allow multiple users without interference. An example is time-division multiple-access or TDMA, which is a widely used method of multiplexing.

The same methods of multiplexing are used for placing a number of systems on an optical fiber. The significant advantage of a fiber-optic system is that the glass fiber is capable of enormous bandwidth. Thus a very large number of users can be placed on a single optical fiber. Also, a large number of copper transmission lines may be replaced with one fiber-optic system, which translates into weight saving. However, the significant advantage of using fiber optics is not so much to save weight or copper, but its extreme immunity to external fields.

SATELLITE COMMUNICATIONS

The majority of communications between ground and air, including data as well as voice, is in the VHF portion of the spectrum. The VHF range is limited to about 200 nmi, and for applications for which the nearest VHF communications facility is more than this distance, alternative forms of communications are required. One of the more important communications systems occupying the VHF spectrum is the Aircraft Communications, Addressing, and Reporting System, or ACARS.

In many cases, HF radio is used as explained earlier in this chapter. HF requires careful selection of frequencies and suffers from a variety of propagation problems and noise. Also, only a limited number of frequencies are available for a large number of users. Listed next are a number of communications requirements for a large aircraft in international travel.

Air traffic control

Company communications for airlines

Emergency communications

Data: pilot reports (pireps), weather information, and the like

Public telephone

Public data modem

Public FAX

This list represents seven types of communications facilities. Multiply this by the number of aircraft flying and the result is a good idea of the amount of communications that is required for the flying population.

One facility for supplying some of the communications services on the list is a communications satellite. A popular communications satellite system is the International Maritime Satellite, or INMARSAT. INMARSAT is an international organization that owns and operates the satellite portion of the INMARSAT system. Countries contribute to INMARSAT through signatories, which invest time and effort in developing the INMARSAT system.

The SATCOM system consists of three parts: the satellites, ground earth stations, GES, and aircraft earth stations, AES. The satellite portion of the SATCOM system consists of three geosynchronous satellites that provide global coverage below 75° of latitude. The geosynchronous orbits have a lack of coverage at the poles, but it must not be forgotten that the satellite system is a maritime satellite.

The ground earth stations are the uplink facilities that provide the connection between terrestrial communications facilities and the satellites. The GES is the link for the satellites to the public switched telephone network, or PSTN.

The aircraft earth stations are the user equipment mounted in aircraft. Like the GES, the AES communicates with the satellite, but the AES is the terminus. All aircraft earth stations must meet all INMARSAT specifications, and before the INMARSAT may be used, authorization must be obtained from INMARSAT.

Although this satellite system was originally instituted for maritime use, it is finding a number of airborne users. One inhibitor to the aeronautical use of the SATCOM system is that the AES uses a high-gain antenna. This requires a large steerable antenna, which for shipboard applications is not a serious detriment, but because of the drag encountered, this high-gain antenna is a problem for aircraft use. However, recent developments in flat phased array antennas for aircraft have made it possible to install low-drag, high-gain antennas on aircraft.

A nonsteered low-gain antenna may be used for the transmission of low-speed data, from 300 to 1000 bits per second. High-speed data and voice require the use of the steerable high-gain antenna.

The satellite-to-AES link is made in the 1500–1600-MHz region of the L band, using frequency-division multiplex, packet communications, and time-division multiplex. The satellite-to-GES link is by radio transmission in the C band at 4 to 6 gigahertz (GHz).

Voice communication is transmitted by 9600-bit/s digital transmission. Public telephone service is available through the PSTN; to place a call, as for any other telephone, the desired number is dialed. Public telephone calls may be made either from ground to air or air to ground.

SATCOM satellites will be a part of a future system called automatic dependent surveillance, or ADS. This system will involve the continuous reporting of position by aircraft, which may be used for flight following, collision avoidance, and so on.

REVIEW QUESTIONS

4.1. What type of modulation is used for VHF communications in aircraft?

4.2. What frequency ranges are used for airborne communications?

4.3. What are some examples of half-duplex communications?

4.4. What is a ground wave? What types of radio services rely on ground waves?

4.5. How does radio energy propagate from a transmitter to a receiver beyond the geometrical horizon?

4.6. How does the ionosphere affect the propagation of radio waves?

4.7. What role do sunspots play in the propagation of radio waves?

4.8. What are the three general requirements of a radio receiver?

4.9. How does the superheterodyne topology solve the problems associated with the TRF radio receiver?

4.10. How much gain, in decibels, is required to amplify a 5-μV signal across 50 Ω to 10 W?

4.11. When is HF radio used for voice communications?

4.12. Why do HF transceivers provide a wide frequency range?

4.13. What is single sideband? Why is it used for high-frequency communications?

4.14. What is crosstalk as applied to an audio control panel?

4.15. Why does the audio panel have an emergency bypass mode?

4.16. What characteristics of ARINC 429 make it more suited for rejecting noise?

4.17. How do some microphones cancel background noise?

4.18. What is the significant disadvantage of the carbon microphone?

4.19. What significant advantages are gained by using fiber optics in an aircraft?

4.20. When must a high-gain antenna be used to communicate through the INMARSAT system?

5

Radio Navigation

CHAPTER OBJECTIVES

In this chapter the student will learn of the historical foundations of radio navigation and some of the criteria that were responsible for the navigation system in use today. The VOR system is covered in detail including DME. The chapter covers the LORAN-C system and the global-coverage Omega system. Satellite navigation is covered in detail.

INTRODUCTION

Early radio navigation systems were based on low-frequency radio transmission. The earliest low-frequency navigation was by radio direction finding; the bearing to a radio transmitter was determined by a directional antenna. The aircraft was turned in the direction of the low-frequency navigation station and the radio transmitter was used as a homing beacon.

Low-frequency radio signals travel by ground-wave propagation, which is desirable for direction finding, but the choice of low-frequency radio waves was also due to the state of technology at the time. The first attempts at radio navigation were in the late 1920s and the radio art was in its infancy. Low-frequency radio direction finding is still being used, but it has its disadvantages, which will be discussed later.

Early direction-finding receivers used a manually rotated antenna that literally pointed to the beacon transmitter. The directional antenna was fitted with a dial that gave the direction to the beacon relative to the heading of the aircraft. Turning the aircraft such that the bearing to the beacon was straight ahead caused the aircraft to fly to the beacon. Improvements in direction-finding receivers produced systems that automatically rotated the antenna and provided a continuous bearing-to-station indication.

The homing beacon provides the heading to the beacon but unless other information is used, the actual position of the aircraft relative to the homing beacon is unknown. As an example, an aircraft flying to a beacon knows that the beacon is straight ahead. But where is ahead? Using the magnetic compass, perhaps ahead is north, which implies that the aircraft is south of the beacon at an unknown distance. In this example, the compass reading is the additional information needed. Figure 5-1 shows three aircraft with the same direction finder indications, but clearly at totally different positions.

A second disadvantage of the homing beacon is evident when the aircraft is flown directly to the beacon by homing. The actual flight path does not necessarily follow a straight-line track as shown in Fig. 5-2. This is because the aircraft may be continually blown

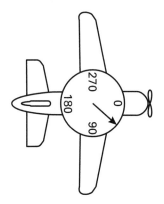

NDB

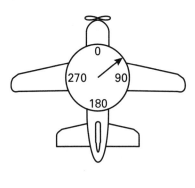

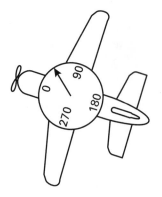

Figure 5–1   Aircraft indications to a homing beacon.

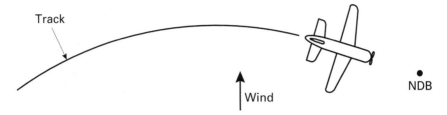

Figure 5–2   Example of a curved track flown by an aircraft because of a crosswind.

off course by prevailing winds. What results is a curved track to the beacon station, although the aircraft maintains a heading to the homing beacon. When the distance is great and there is a moderate crosswind, the deviation from the desired course could be significant. However, maintaining a crab angle, the angle between the heading of the aircraft and the actual course, based on the prevailing winds would permit a straight-line track. But this involves calculations and navigation techniques external to the radio navigation and increases the pilot workload.

The nondirectional beacon (NDB) provides heading information, but what was needed in the early days of radio navigation was a navigation aid that provided course information, essentially an electronic line in the sky that could be flown. The method undertaken to provide this type of navigation was the four-course A–N range. The four-course range provided four quadrants of radio signals by using two transmitters and two antenna arrays that are superimposed. In two of the quadrants, the Morse code letter A was transmitted and in the other quadrants the letter N was transmitted, as shown in Fig. 5-3. At the areas where the quadrants intersect, the letters A and N merge to provide a steady tone. To fly a range, the aircraft is flown until the A or N heard in the speaker or headphones merges to a steady tone. This indicates that the aircraft is on one of the four courses available. The

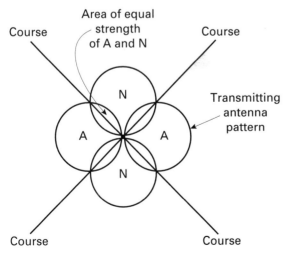

Figure 5–3   Radiation pattern of a four-course A–N range.

aircraft is kept on course by making corrective maneuvers to restore the steady tone. The major disadvantage of this form of navigation aid is that only four courses from the transmitter are available for navigation, and this range system has long been obsolete.

There are common problems with all low-frequency navigation systems. The first is low-frequency atmospheric noise caused by weather conditions. A second problem is unpredictable propagation. In spite of these problems, the homing beacon, or the NDB, is used for local navigation throughout the world. The basic attraction of the NDB is the very low cost of the ground facilities.

A very important requirement for the radiation of an NDB for radio direction-finding use is that the signal propagation be by ground wave only. Signals propagated by the ionosphere are so severely distorted as to make their use for radio direction-finding undesirable. Thus, nondirectional beacon transmitters are vertically polarized, which is the only polarization that supports ground-wave propagation.

DIRECTION-FINDING RECEIVERS

The direction-finding receiver, whether automatic or manual, uses a directional antenna to determine the bearing to the NDB. All low-frequency direction-finding receivers employ magnetic field antennas, which are, essentially, a coil of wire. The magnetic field lines of flux pass through the coil and induce a current in the loop.

The magnetic field vectors are arranged such that they are tangential to concentric circles, with the NDB transmitter at the center of the circles. The maximum amount of magnetic lines of flux are intercepted when the axis of the loop antenna is aligned with a line drawn to the transmitting station, as shown in Fig. 5-4. The number of magnetic flux lines passing through the loop is proportional to the sine of the angle between the normal to the plane of the loop and the radius to the NDB antenna.

If we were to plot the signal level as a function of the angle between the perpendicular of the plane of the loop and the direction to the transmitter, the resulting plot is called an *antenna pattern* and is shown in Fig. 5-5.

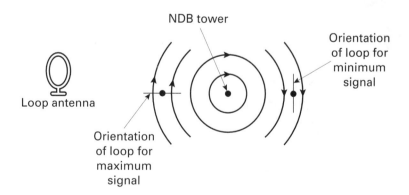

Figure 5–4   Magnetic field orientation of a NDB transmitter and the loop antenna.

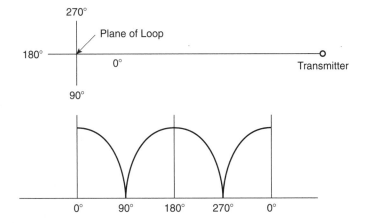

Figure 5–5   Signal strength of a received signal as a function of the relative angle.

The amplitude of the voltage induced in the coil is lowest, or at a null, when the plane of the loop is perpendicular to the bearing to the transmitter. Either the peak or the null may be used to determine the bearing to the navigation beacon. It is easier to see the sharp null on a carrier strength meter and a more accurate determination of the null is available. However, when the level of precipitation static is high, this would tend to obscure the signal and make it difficult to find the null. Therefore, the antenna is adjusted for a peak during conditions of high static. The difference between the peak and the null is 90° and this must be taken into account.

The antenna pattern shown in Fig. 5-6 is based on the fact that the voltage induced in the loop antenna is due to magnetic field lines only. The radiation of radio energy involves both magnetic and electric fields, and the electric fields can induce voltage in conductors. If the electric field is capable of inducing voltage in the loop antenna, the antenna pattern of Fig. 5-6 will be distorted. The loop antenna is prevented from picking up energy from the electric field by shielding the loop from electric fields. To shield it from electric fields, the antenna is encased in a conductive housing. In the case of a loop antenna, the wire comprising the loop is placed in a conductive pipe such as copper or aluminum.

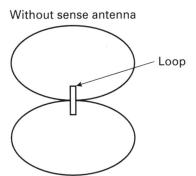

Figure 5–6   Antenna pattern of a loop antenna.

The loop antenna and the shield are similar to a transformer, with the shield acting as a secondary. Recalling the transformation of impedance by a transformer, if the secondary is shorted, regardless of the turns ratio, the zero impedance will be reflected to the primary. In this case the primary is the loop antenna and a zero impedance will be reflected to the primary and cause significant signal loss. To prevent this, the shield is separated by a thin cut to prevent a continuous circuit but provide electric field shielding.

Whether a null or a peak is used for direction finding, there are two antenna orientations that will provide either a null or peak, which results in an ambiguity. The magnetic field antenna is used in conjunction with an electric field antenna to remove the ambiguity and to provide a single peak or null.

When the magnetic field antenna goes through a null, the polarity of the induced voltage reverses phase at the null. This is because the magnetic field lines begin to enter from the other side of the loop and thus induce an opposite polarity current. Therefore, there is a significant difference between the two peaks in that the phase of the induced voltage is the opposite of the other peak.

If there is a phase reference for the carrier, the polarity of the loop antenna output can be determined and the two peaks can be differentiated. This phase reference is obtained from the electric field of the NDB signal. A simple, short wire antenna provides an electric field reference and is called the *sense antenna*. There is a 90° electrical phase shift between the electric and magnetic fields, so a 90° phase shift is inserted after the electric field antenna. Therefore, the phase-shifted output from the electric field antenna is in phase with one of the peaks of the magnetic field antenna and out of phase with the other peak. The amplitude of the electric field antenna output is adjusted so that it is exactly equal to the maximum output from the magnetic field antenna. Thus, when the magnetic field antenna is rotated such that the maximum output voltage is equal in amplitude but out of phase with the voltage from the electric field antenna, there is complete cancellation, creating a null as shown in Fig. 5-7. When the antenna is oriented so that the magnetic and electric field components are in phase, the resulting amplitude is twice that obtained with just the loop alone and causes a peak. Thus, the combination of loop and sense antenna has one peak and one null.

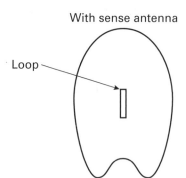

With sense antenna

Loop

Figure 5–7   The combination of the electric field and magnetic field creates an antenna pattern with one null and peak.

If the polarity of the loop is changed, the antenna pattern will change by 180° degrees, because what was added in phase is now out of phase and what was out of phase is now in phase.

A large loop antenna on the skin of an aircraft represents considerable drag. One method of reducing the size of an ADF antenna without reducing the performance of the antenna is to fill the area in the center of the loop with a ferromagnetic material. As explained in Chapter 2, ferromagnetic materials, particularly those made from ferrites, have an ability to concentrate magnetic field lines. An antenna made with a ferromagnetic core can be made with an area of the loop that is 1/100th of an air-core loop and will still have the same level of performance.

Even with ferromagnetic material, we still have an antenna that must be rotated for direction finding, which is undesirable. Although the ferromagnetic core has reduced the size of the loop, a motor is required to turn the antenna, which must be on the outside of the fuselage. This means that the motor and its ancillary drive mechanism are exposed to the full temperature variations outside the aircraft. As undesirable as this is, ADF antennas were made this way for a number of years. What is needed is an antenna that has no moving parts, can have a small profile, and can be more robust.

One method of removing the rotating antenna from the exterior of the aircraft is to sample the magnetic field outside the aircraft and re-create that field within the aircraft, where a miniature rotating antenna can be placed. Although there are still moving parts, they are inside the aircraft and not subject to the temperature and pressure variations found on the outside of the aircraft skin.

The technique of achieving this is to mount two fixed loops at 90° to each other on the outside of the aircraft and to conduct the received signals to the instrument panel, where a rotating loop antenna may be employed. This device is called a *goniometer*. This is usually accomplished by winding two separate orthogonal windings on a block of ferrite, as shown in Fig. 5-8, where each loop feeds a second loop inside the aircraft.

It is important to remember that the magnetic field is a vector. Imagine a coordinate system in which one pickup loop is oriented parallel to the X axis and the other loop, being orthogonal, is oriented parallel to the Y axis. As previously discussed, the voltage induced in the loop is proportional to the sine of the angle between the normal to the plane of the loop and the vector to the transmitting antenna. Therefore, the voltage induced in one loop is proportional to the X component of the magnetic field vector, while the other pickup loop output is proportional to the Y component.

These two currents, proportional to the X and Y components of the magnetic field vector, are passed through two coils that are oriented with the X and Y axis of another coordinate system, in this case, inside the aircraft. The result is the generation of a magnetic field vector with exactly the same orientation as the outside magnetic field but relative to the inside coordinate system. The magnitude of the vector re-created may not be the same, but since we are concerned with peaks and nulls for direction finding, this does not matter.

A small loop, called the *rotor*, which is a miniature of the simple rotating loop used in early direction finders, is placed in the magnetic field within the goniometer. The signal from this miniature loop behaves exactly as an external rotating loop antenna. Thus, the loop

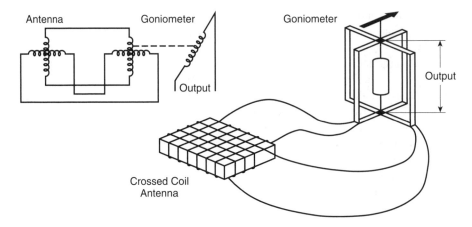

Figure 5–8   Goniometer.

may be rotated from within the aircraft, and only a small nonmoving antenna is placed out-side. Like the outside loop, the goniometer loop antenna must be shielded from external electric fields. This is usually done by shielding the entire goniometer. A pointer is mounted to the goniometer rotor, and the loop voltage is added to the phase-shifted sense antenna output to provide a system with a single peak and null.

AUTOMATIC DIRECTION FINDING

The manual direction-finding receiver is a simple superheterodyne that provides an output that is simply a function of the received signal strength. The receiver is also required to provide an audio output for the purpose of listening to the Morse code identifier applied to the NDB carrier or voice broadcasts. For the navigation function, only the relative signal strength is required to find peaks and nulls.

The direction-finding receiver is a conventional superheterodyne, with the exception that the input must sum the loop and sense antennas with the necessary 90° phase shift. If a meter is applied to the signal strength output of the receiver, the goniometer rotor can be rotated manually for a peak or null, and the goniometer pointer will indicate the relative bearing to the NDB.

To make the direction finder fully automatic, a motor is connected to the rotor of the goniometer and controlled such that the goniometer is driven to either the signal null or peak. What is needed to do this is some sort of left–right signal that may be used to turn the motor in the correct direction.

The phase reference from the electric field antenna can be used to provide a left–right

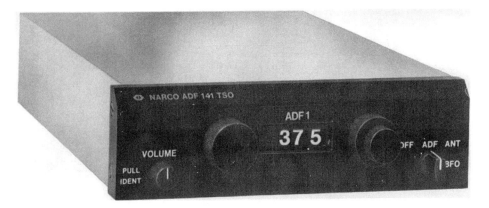

Photo 5–1   An ADF receiver for light general aviation aircraft. Photo courtesy of Narco Avionics, Inc.

reference at the peak or null. But the signals at the sense antenna and the loop are too weak and corrupted with other, interfering signals to be used directly with a phase detector.

The electric field antenna signal and the loop signal can be processed in separate receivers where the interfering signals are removed and the signals are raised to a level that can be used with a phase detector to generate the left–right signal. The problem that makes this technique of using two receivers virtually impossible is that the phase shift through the two receivers would have to be matched to prevent the generation of serious errors.

An alternative method is to process both the sense and loop signals through the same receiver. This assures identical phase shifts and reduces the complexity of the automatic direction finder by requiring only one receiver. This is accomplished by modulating the output of the loop antenna using double-sideband suppressed-carrier and combining the modulated signal with the sense antenna after the 90° phase shift. The result is a full-carrier amplitude-modulated signal in which the sidebands are from the loop and the carrier is from the sense antenna; the signal can be processed in a normal fashion with a superheterodyne receiver, as shown in Fig. 5-9.

The loop antenna is modulated using a balanced mixer, which is like a reversing switch. When the modulating signal applied to the balanced modulator is positive, the output is the same phase as the input. When the input to the modulator is negative, the phase is reversed.

Remember, when the loop and sense antenna are added, a directional antenna pattern results with only one peak and one null. If the phase of the loop is reversed, the direction of the peak and null will reverse.

The signal at the output of the balanced mixer is essentially switching between two antenna orientations, as shown in Fig. 5-10. When viewed as a function of time, the signal represents a full-carrier double-sideband AM signal with square-wave modulation. The carrier is from the sense antenna and is unaffected by the modulation.

The relative amplitudes of the two halves of the square-wave modulation depend on where the NDB is relative to the switched antenna patterns. As shown in Fig. 5-11, if the

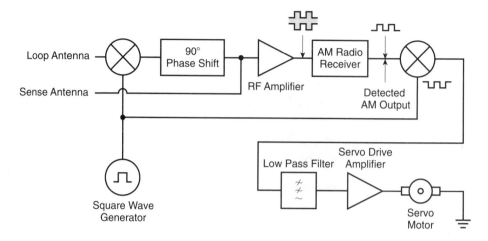

Figure 5–9   Block diagram of an ADF receiver.

NDB is in the direction of one of the two antenna orientations, that part of the square-wave modulation has the greatest amplitude. If the NDB lies in the direction of the other antenna orientation, the amplitude of the balanced mixer is greater in the other half of the modulation cycle. If the NDB lies in the center of the two antenna orientations, both halves of the modulation cycle will have the same amplitude, and the output of the balanced mixer will have no amplitude modulation.

The amplitude-modulated signal is amplified and demodulated through the conventional superheterodyne receiver. The demodulated output of the receiver is compared with the original modulation applied to the loop signal.

The comparison is done with another balanced modulator, which, as with the loop signal, is a reversing switch. The first half of the modulation cycle is amplified with a pos-

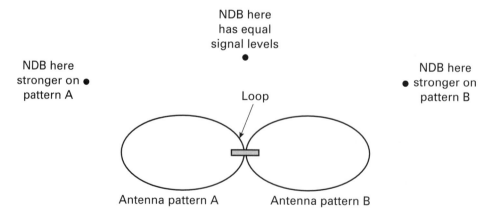

Figure 5–10   The switch of two antenna patterns produces the needed left–right information for an ADF receiver.

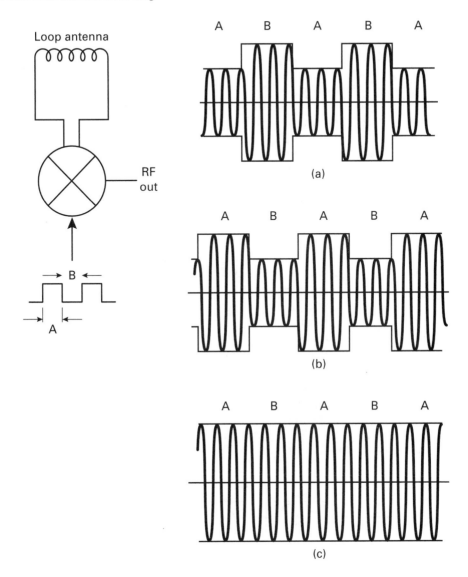

Figure 5–11 Output of the balanced modulator in a servo-type ADF receiver.

itive gain, whereas the other half-cycle is amplified with negative gain. Therefore, if the NDB is in the direction of the antenna orientation during the first half-cycle of the modulation, the demodulated envelope is stronger during the first half-cycle and the output will be a positive voltage. If the NDB is in the direction of the antenna orientation during the second half-cycle, the output will be a negative voltage.

If the NDB orientation is exactly midway between the two antenna orientations, the result will be zero voltage, because amplitude of the envelope is the same for both halves of the modulating cycle. Figure 5-11 shows the output of the balanced modulator as a function of NDB orientation relative to the two antenna patterns.

The demodulated output, which is a square wave, is changed to a dc voltage for which the amplitude is proportional to the square-wave amplitude and the polarity of the dc is a function of the location of The NDB. This dc voltage is used to drive a motor that drives the goniometer coil, with the indicator pointer attached, to a null. This technique of mechanically rotating a shaft automatically from electrical signals is called a *servo system* and thus this is called a *servo-type* automatic direction finder.

The modulation applied to the loop output is at a very low frequency, well below the speech band, so the modulation applied within the ADF receiver can be filtered out so that it will not interfere with the voice modulation and Morse code identifier applied to the NDB signal. Likewise, the modulation applied to the NDB will not interfere with the direction-finding circuits of the receiver.

All-electronic ADF Systems

The ADF still has a sensitive mechanical component, the goniometer. An improvement in ADF cost and reliability could be made if the goniometer were eliminated. The block diagram of a receiver that requires no goniometer is shown in Fig. 5-12. In this example, no goniometer is used and the two loops, the X and Y loops, which are sometimes called the sine and cosine loops, are modulated with two low-frequency square waves. The important characteristic of these two modulating signals is that they are exactly the same frequency but are different in phase by 90°. An interesting characteristic of signals that differ by 90°, which are said to be *in quadrature,* is that they may be processed together in a radio receiver and demodulated separately. The two loops are modulated in quadrature, which produces two double-sideband suppressed-carrier signals that are combined with the sense antenna as with the servo-type ADF. The result is a full-carrier amplitude-modulated signal that has two sets of sidebands in quadrature. The combined signal is processed with a superhetero-

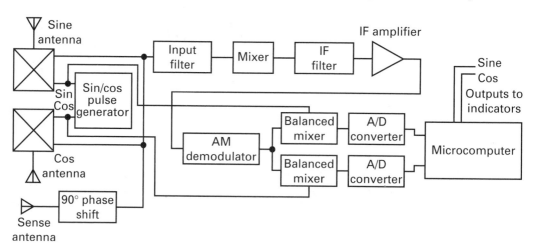

Figure 5–12   Pointing type of ADF.

dyne receiver. By using a quadrature demodulator, which is phase locked to the original modulating signal, two dc outputs result; one is proportional to the amplitude, in both sign and magnitude, of the X loop magnitude and the other is proportional to the Y loop magnitude. The results are two dc signals that represent the magnitude of the X and Y components of the magnetic field vector of the NDB station. There can be used to drive a mechanical pointer as a conventional ADF-style indicator or may be electronically displayed. Even though a mechanical pointer may be involved, this pointer is not a part of a goniometer. When the display is electronic, there are no moving parts.

The ADF receiver provides an audio output that is used to monitor the Morse code identity and any voice modulation. The nondirectional beacons are modulated with a 1000 hertz (Hz) tone to provide the Morse code transmitter identification.

In some countries, primarily in Europe, the NDB identification is provided by turning the carrier on and off with the Morse code. Since an unmodulated carrier produces no audio from a receiver, some method of making the carrier audible is necessary. The technique used is called a *beat frequency oscillator;* it provides an audio-frequency tone when the carrier is present.

ADF Systems and Installations

The performance of an ADF system requires the correct installation of the individual parts of the system. The typical components of an ADF system are the receiver, antenna, and indicator. If the installation is a remote-mounted receiver, a control head must be included. In some cases a separate sense and loop antenna assembly is used.

If the ADF system is a servo-type system, the indicator is usually a goniometer that includes a special interconnecting cable. This cable may not be cut to decrease or increase its length without consulting the manufacturer's instructions. The loop antenna is an inductor, and the effects of the capacitance of the interconnecting cable from the goniometer to the antenna are considered in the calibration of the ADF system.

Another important factor that must be considered is the location of the ADF antenna. The antenna is either specifically made to mount on the top or bottom of the fuselage, or the antenna can be switched by moving wires on a terminal board. If the antenna is incorrectly mounted, the ADF system will not point accurately.

The ADF antenna is best installed as far as possible from large metallic objects such as gear or stabilizers. The effects of these objects cause an error called *quadrantal error.* The largest generator of quadrantal error is the aircraft fuselage. Because the fuselage is metallic, the magnetic field lines are distorted because of currents in the fuselage. The nature of the distortion depends on whether the aircraft fuselage is parallel or perpendicular to the magnetic field lines. This distortion produces a bearing error that is the same but of opposite sign for bearings that differ by 180°.

Because a conducting fuselage is something practically every aircraft has, each aircraft has some quadrantal error. Therefore, ADF receivers have quadrantal-error-correction circuits. The only way quadrantal error can be corrected is to measure the amount of error with the ADF system installed in the aircraft. This is done using a technique called *swinging* the aircraft.

The ADF receiver is tuned to an NDB at a known direction with the aircraft on the ground. The aircraft is then rotated about the full 360° while the error is recorded. The quadrantal error is corrected by turning screwdriver adjustments or by programming through switches. Older receivers, particularly the servo types, had a mechanical correction device that required screwdriver adjustments. The newer all-electronic ADF systems can store quadrantal error correction in a nonvolatile computer memory.

It is not necessary to swing all aircraft to correct for quadrantal errors. If an identical ADF installation is made in other aircraft of the same type, the quadrantal-error-correction factors may be used. Therefore, it is not necessary to swing every aircraft that receives the ADF installation. This is particularly important in air-transport-type aircraft for which the cost of such an exercise could be significant.

THE OMNIRANGE

The VHF omnirange or VOR was invented in the early 1940s to solve a number of problems associated with homing beacons and the four-course range. The first problem to be solved was the unreliable operation of the low-frequency beacons and ranges due to precipitation static and radio propagation effects. The new VOR operated in the VHF part of the radio spectrum where there is practically no precipitation static and where the radio signals travel in essentially straight lines.

The second problem was the limited navigation capability of the low-frequency beacons and ranges. The new VOR system offered the ability to select any desired bearing and to fly to or from the VOR station on what is called a *radial*. If the pilot wishes to fly to the VOR on a specific radial, the desired radial is set using the omni bearing selector or OBS. The VOR equipment will indicate the deviation from the desired course and that flying the desired heading will take the aircraft TO the VOR, as shown in Fig. 5-13. If the VOR indicator indicates FROM, then flying the same heading as the OBS will take the aircraft farther away from the VOR. When the aircraft passes over the VOR ground station, the TO/FROM indicator changes, signifying that the aircraft has reached the VOR. The VOR receiver provides steering information to keep the aircraft on the desired radial, which is called the *course deviation indicator.* Therefore, the VOR indicator has three components, the omni bearing selector or OBS, the course deviation indicator or CDI, and a TO/FROM indicator.

Figure 5-14 shows the state of an omni indicator for various locations around a VOR station. It is important to understand that, unlike the ADF, the heading of the aircraft has no effect on the omni indication.

There are 160 frequencies designated as VOR frequencies in the range of 108.00 to 117.95 megahertz (MHz). The channels are separated by 50 kilohertz (kHz). There are 40 frequencies that are not used for VOR operation placed at the odd 100 kHz and odd 100 kHz plus 50 kHz below 112 MHz. These frequencies are reserved for landing systems.

The VOR system operates using a rotating antenna pattern. The VOR rotates a directional antenna pattern at 30 revolutions per second. The antenna does not physically rotate, but the antenna pattern rotation is accomplished electronically. Rotating a directional radio antenna is similar to a rotating airport beacon. An observer sees the white light followed by the green light, then the white light, and so on. If the observer knows when the white

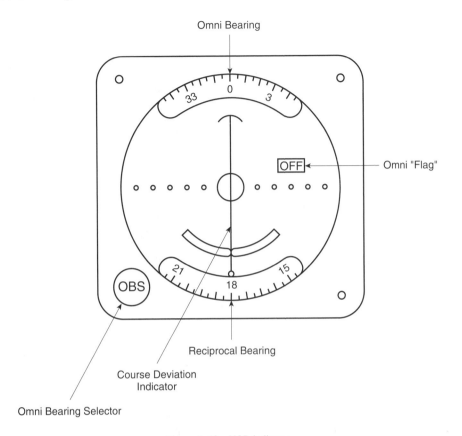

Omni Bearing

Omni "Flag"

Reciprocal Bearing

Course Deviation
Indicator

Omni Bearing Selector

Figure 5–13   VOR indicator.

light of the beacon points north and the rate at which the beacon turns, the observer's angular bearing from the beacon can be determined by measuring the time from when the light points north to when the light points at the observer. The rate of rotation can be determined by measuring the elapsed time between green or white flashes. Imagine that a second light is provided, but this light has an omnidirectional coverage so that it can be seen anywhere around the beacon tower. This omnidirectional blinking light provides a pulse of light when the beacon's white light is pointing magnetic north. Also, assume that the beacon rotates exactly once every 6 seconds, or 60° per second. The angular position relative to the beacon tower can be easily determined by measuring the time delay from the blinking of the omnidirectional light to the passing of the rotating light and multiplying that time by 60°. The VOR system achieves the same situation electronically.

In the VOR system, the rotating antenna pattern causes an amplitude-modulated RF carrier in which the amplitude rises and falls as if it were modulated with a 30-Hz sine wave. What is needed is a method of providing a reference signal that will allow determining when the rotating antenna pattern is pointing north. A 9960-Hz subcarrier is applied to the RF carrier of the VOR transmitter by amplitude modulating the carrier.

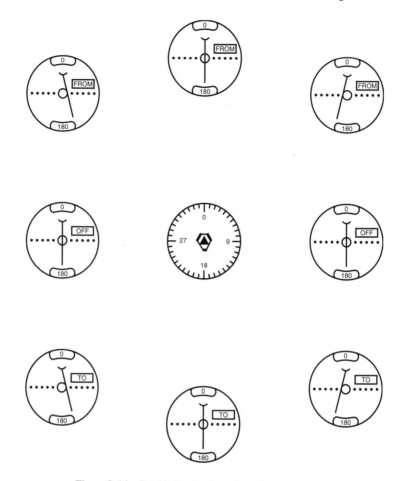

Figure 5–14 Omni indication for various aircraft positions.

The need for a carrier was explained in Chapter 2. A subcarrier also transmits information by applying modulation to a carrier, except that the subcarrier is modulating a main carrier. Subcarriers are used to supply more than one information signal to a main carrier. In the case of the VOR, the subcarrier provides an independent method of supplying the reference signal.

The subcarrier is frequency modulated using a 30-Hz sine wave. It is necessary that this second sine-wave modulation not only be exactly the same frequency, but that the phase angle be equal to the VOR bearing. This second sine wave is the reference against which the phase of the variable 30-Hz modulation will be compared.

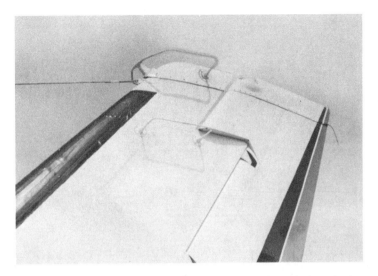

Photo 5–2   The tail of a medium-sized twin-engined aircraft showing a navigation antenna referred to as a "towel bar" because of its similarity to a common bathroom fixture. An ADF sense antenna is also anchored to the aircraft tail.

THE DOPPLER VOR

The VOR system was invented in the 1940s and the basic scheme was developed around the technology available at the time. The system has some basic weaknesses that have been improved through better hardware over the years.

One weakness of the VOR system is signal corruption due to multipath signal propagation. Multipath is the mortal enemy of all radio transmissions and has been discussed at various points in this text. The effects of ionospheric multipath on signal propagation at the VHF frequency range are much less than for the low-frequency radio navigation aids the VOR replaced, but multipath signal propagation is present in any type of radio transmission.

The use of a directional antenna is the main cause of multipath distortion in the VOR system. If there is a reflecting object such as a mountain in the beam of the antenna, it could reflect the VOR signal. When the antenna is pointing away from the receiver, the signal should be at the minimum of the 30-Hz amplitude modulation. Because the receiver can receive a reflected signal that combines with the direct signal, the actual peaks and nulls of the modulation envelope are distorted. This can cause severe error in the measured VOR bearings.

At some locations a VOR of the conventional type cannot be installed because there are significant reflecting objects such as mountains. Problems associated with the surrounding terrain or structures are called siting problems, which refers to the difficulties associated with the VOR site.

A method of reducing the effects of multipath is to use nondirectional antennas. The greatest distortion of the received signal occurs when the directional antenna is pointing at

Photo 5–3   A complete radio navigation and communications package including audio panel, VOR, ILS, ADF, and indicators. Photo courtesy of Terra Corporation.

the reflector and away from the receiver. This situation does not occur when using non-directional antennas. The use of nondirectional antennas does not eliminate multipath distortion but it is an improvement.

Because nondirectional antennas are used, a different technique must be used to generate the VOR signal. Rather than a rotating antenna pattern, moving antennas are employed. Just as the rotating antenna pattern does not employ a mechanically rotating antenna, the moving antennas are electronically generated.

The movement of the antenna is around a circular path, as shown in Fig. 5-15. When an emitter of a periodic signal is moving relative to a receiver, a frequency shift occurs that is a function of the relative velocity; this is Doppler shift. This shift is the familiar change in pitch of an automobile horn as the vehicle passes by. In the Doppler VOR system, the movement of the antennas causes a 480-Hz Doppler shift, which produces frequency modulation. With the moving antennas making one complete rotation every 1/30th of a second, the frequency modulation is at a 30-Hz rate. As the antenna is moving toward the aircraft, the frequency is increased, and when the antenna is moving away from the aircraft, the frequency is decreased. The Doppler VOR uses two moving antennas, one 9960 Hz above and a second 9960 Hz below the nominal carrier frequency. The antennas are located on opposite sides of the circular track around which the antennas move. Therefore, while one antenna is moving toward the receiver, the other is moving away. Remember, the antennas on the track do not move. What the Doppler VOR station has is a large number of nondirectional antennas in a phased array; each antenna is fed a specific amplitude and phase relationship so that the antenna can be electronically moved around the circular track.

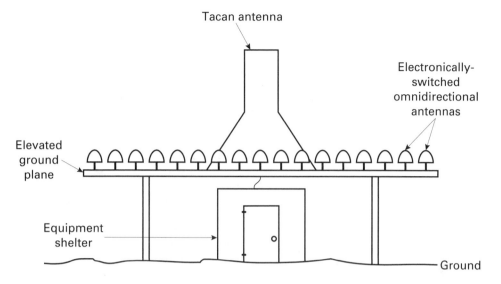

Figure 5–15   Construction of a Doppler VOR station.

The amount of frequency modulation is a function of the velocity along a line from the VOR station and the aircraft. Because the motion is in a circle, the maximum velocity toward the aircraft occurs when the antenna is located at 90° relative to the aircraft's bearing. The minimum Doppler shift occurs when the antenna is on the line to the aircraft.

The relative velocity is proportional to the sine of the angle between the aircraft's bearing from the VOR and the position of the antenna.

Example

What is the required radius of the Doppler VOR antenna path for a VOR facility operating at 110 MHz to produce peak frequency modulation of 480 Hz?

Solution.   The peripheral velocity is $V = 2\pi R(30)$, where R is the radius of the circular path and the factor 30 is due to the fact that the path is covered 30 times a second. Of course, $2\pi R$ is the circumference. The peripheral velocity is directly toward the aircraft at the 90° and directly away at the 270° points and thus produces the peak deviation of 480 Hz at these points.

Recall from physics that the Doppler shift is $f = VF/c$, where V is the velocity in the direction of the receiver, c is the speed of light, and F is the carrier frequency. Therefore,

$$f = \frac{2\pi R(30) * 110 \times 10^{6}}{c} = 480 \text{ Hz}$$

Solving for R,

$$R = \frac{480 \text{ Hz} * 3 \times 10^{8}}{2\pi(30) * 110 \times 10^{6}} = 6.9 \text{ meters (m)}$$

Photo 5–4   Two different navigation receivers and two indicators. One indicator has two pointers for use with a glide-scope system. Photo courtesy of Narco Avionics, Inc.

An omnidirectional antenna is also provided and amplitude modulated with a 30-Hz waveform. The VOR receiver receives three signals; the 30-Hz amplitude-modulated signal at the nominal carrier frequency; one signal 9960 Hz below the nominal carrier frequency, which is frequency modulated at 30 Hz; and a signal 9960 Hz above the nominal carrier frequency, which is also frequency modulated with a 30-Hz sine function. The two 9960-Hz offset signals appear as sidebands as if a nominal carrier frequency were modulated with a 9960-Hz subcarrier. The two sidebands both have frequency modulation such that, as the instantaneous frequency of the lower sideband is decreasing, the upper sideband is increasing. This is the equivalent of frequency modulating the 9960-Hz subcarrier.

There are two differences, however, between the received Doppler VOR signal and a conventional VOR. First, the radiation of all three antennas is from omnidirectional antennas and less prone to multipath distortion. The second is that the phase of the 9960-Hz subcarrier frequency modulation is a function of aircraft position and not of the 30-Hz amplitude modulation. This is the opposite of what is encountered in conventional VOR. However, VOR receivers perform a phase-angle measurement, and it makes no difference whether the reference or variable changes phase relative to the aircraft's position.

Doppler VOR stations are considerably more expensive than conventional VORs and have been installed where there are known problems with multipath. In spite of the cost, Doppler VOR ground stations have been installed in large numbers around the world to improve VOR operation in areas with severe siting problems.

The VOR Receiver

Figure 5-16 shows the block diagram of a VOR receiver and navigation converter. The VOR receiver is a conventional single-conversion superheterodyne receiver using a frequency synthesizer for the local oscillator. This receiver is essentially the same as the receiver outlined in Chapter 4 used for VHF communications. There are some differences due to the nature of the navigation signal. Usually, one receiver is used for both VOR and localizer reception and thus the receiver provides coverage from 108.00 to 117.95 MHz.

The IF amplifier feeds a simple amplitude modulation detector that provides the necessary feedback for the receiver automatic gain control (AGC) and the demodulated audio signal, which is called *nav composite*. This audio output is fed to a navigation signal converter, rather than directly to the loudspeaker as would be the case with a communications receiver.

It is necessary to separate the reference and the variable 30-Hz sine waves and measure the phase angle between the two to determine the VOR radial.

The nav composite is passed through a high- and a low-pass filter. The low-pass filter separates out the 30-Hz variable waveform. The output of the high-pass filter is the 9960-Hz subcarrier, which contains the 30-Hz reference.

To remove the 30-Hz frequency modulation from the 9960-Hz subcarrier, the 9960-Hz output from the high-pass filter is fed to a limiting amplifier. A limiting amplifier is an

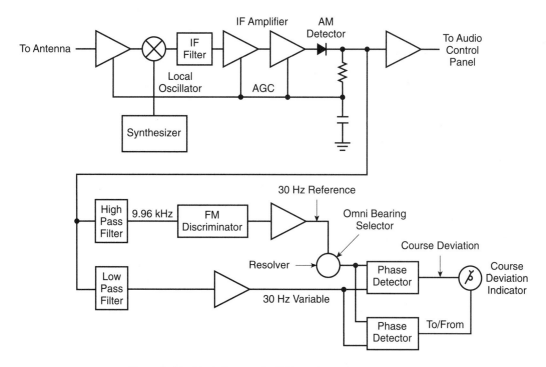

Figure 5–16   Block diagram of a VOR receiver and NAV converter.

amplifier that is driven to saturation so that a change in input amplitude will not produce a significant change in output amplitude. The output of the limiting amplifier is fed to a frequency discriminator, which removes the 30-Hz reference modulation.

The variable 30-Hz waveform is phase shifted with a circuit that uses a resolver. A *resolver* is a special form of motor with two fixed coils and a rotating coil. The resolver is one of the specialized motors that is not used to move things but for precision position sensing and control. These motors are made with great precision and high-quality components for low-friction accurate positioning. The resolver is used with a circuit that converts the mechanical shaft position to an equivalent phase shift. The resolver is connected to a dial, which is the OBS. The output of the OBS phase shifter is fed to a phase detector along with the reference sine wave. The phase detector drives the CDI meter arranged so that the CDI reads zero when the reference and the phase-shifted output of the OBS are in phase. This occurs when the bearing as selected by the OBS is equal to the received VOR signal's bearing. Should the received signal deviate from the bearing selected, the CDI will move from the zero position and indicate the required correction.

If the aircraft is flying in the same direction as the TO/FROM indicator reads, the CDI points to where the selected radial is located. That is, if the aircraft is flying to a VOR on a TO radial and the aircraft is to the right of the selected radial, the CDI points to the left, indicating that the selected radial is to the left. To fly a VOR radial, the pilot flies into the meter, that is, flys to the left when the CDI points left and to the right when the CDI points right. It must be understood that this is only true when the aircraft is flying to or from the VOR as indicated by the TO/FROM indicator. If this is not the case, the CDI indications are reversed.

The TO/FROM indicator is derived from a second phase detector that operates on the same reference and phase-shifted variable signals from the OBS output, except that the phase detector senses a 90° phase angle. Therefore, the output of the phase detector is a function of the position of the aircraft relative to an omni radial that is 90° from the OBS radial. Referring to Fig. 5-14, it can be seen that this 90° offset radial is the dividing line between the TO and FROM halves of the omni coverage.

Signal Integrity Monitoring

It is important for safe navigation that all radio navigation signals be continually tested for signal integrity. A number of monitoring facilities are associated with every VOR transmitter. These include monitors contained within the transmitter, as well as receivers mounted a short distance from the transmitter site. Should a failure of the VOR occur such that the accuracy of the transmitted signal is degraded, the facility would be shut down so that the aircraft could not navigate with the corrupted signal. The integrity of the received signal is also monitored in the airborne VOR receiver. Even with a good transmitted VOR signal, corruptions to the received signal can occur due to interference from another VOR, a very weak signal, or a failure of the VOR receiver.

To prevent navigation based on a corrupted signal, the VOR receiver contains circuits for evaluating the received signal. The level of the 30-Hz reference and the 30-Hz variable are sensed along with the received signal strength. If the three parameters are not sufficient to perform reliable navigation, a detector will display a flag, which warns the pilot that the signal is not trustworthy.

In addition to the navigation signals, the VOR is modulated with a Morse code identifier and, in some cases, by voice. Prudent navigation procedures require the pilot to listen to the Morse code identifier to ensure that the navigation station is transmitting and that navigation is to the correct VOR. A voice transmission is available for a number of purposes, including weather information and ground to air communication.

Signal integrity monitoring of a nondirectional beacon involves the ground station monitoring its own signals with nearby receivers, as with the VOR. In the case of direction finding, the transmitted signal required for navigation is only an unmodulated carrier. The Morse code ident modulation serves as the signal integrity monitor for the airborne component. The pilot listens to the ident, identifies the station, and determines that the signal is clear and not distorted.

The VOR system is the basis for specific air routes called *victor* airways. Victor is the phonetic word for the letter V, which refers to VOR. The VOR system is used to define specific routes that begin and terminate with a VOR.

DISTANCE MEASURING

The use of VOR allows the aircraft to be flown on a specific path called a radial to or from a VOR station. The addition of the distance from the VOR to the aircraft places the aircraft at a specific point using a polar coordinate system with the VOR station at the origin. The aircraft position will be stated in terms of a radial and a distance relative to the VOR. What is needed is a radio-based method of determining the distance to the VOR ground facility.

In the sixteenth century, Galileo attempted to measure the speed of light by sending his assistant to a distant mountain top with a lantern. The plan was for Galileo to uncover a lantern and when the assistant saw the light, the assistant would uncover his lantern. The elapsed time from when Galileo uncovered his lantern until he saw the light from the assistant's lantern would be the time required for the light to make the round trip from one mountain top to another and return, plus the reaction time of the assistant. Galileo had no concept of the velocity of light and essentially saw no delay other than the reaction time. He correctly deduced from his experiment that the velocity of light is very great.

Galileo was attempting to measure the speed of light by knowing the distance between the mountains, but if we know the speed of light, we can measure the distance between the mountains.

The measurement of distance using radio waves uses Galileo's scheme, except with accurate high-speed clocks to measure the elapsed time and a fast electronic "uncovering" of the energy source. The system used for measuring distance with radio is called *distance-measuring equipment* (DME) and can be used as an adjunct to the VOR and instrument landing system.

A radio transmitter in the aircraft transmits two short pulses of radio energy, called an *interrogation,* which are received by a *transponder* at the VOR ground facility. After a fixed delay time, which could be viewed as Galileo's assistant's reaction time, the transponder replies to the interrogation. To determine the distance, the airborne equipment measures the time delay from the transmission of the two pulses and the receipt of the reply and subtracts the reaction time delay. This produces the time it took to traverse the path from the aircraft to the DME transponder and back. The distance from the aircraft to the DME transponder is given by the formula

$$D = \frac{(t-r)}{2}\, c$$

where D is the distance in meters, t is the delay time in seconds, r is the reaction time in seconds, and c is the velocity of light.

The concept of the *radar mile* is used to simplify the distance equation. A radar mile is not a distance but the time required for a radio signal to travel 1 nautical mile (nmi) and return, or a total distance of 2 nmi, which is 12.35 microseconds (μs). Using this concept, the distance equation is rewritten as

$$D = \frac{t-r}{12.35\ \mu s}$$

where D is now in nautical miles.

This distance, D, is the actual distance from the aircraft to the VOR and is sometimes called *slant range*. The slant range is the hypotenuse of a triangle where the map distance represents one side of a right triangle and the aircraft's altitude represents another, as shown in Fig. 5-17. The third side is the actual distance along the ground, assuming a flat Earth. This is a safe assumption because the distances encountered with distance measuring are a hundred nautical miles or so compared with the circumference of Earth of 21,600 nmi. The differences between the actual distance along the ground, which is the same as shown on a map, and the distance calculated, assuming a flat Earth, is very small.

When the aircraft is a long distance from the DME transponder, the slant range and map range are very close. For high-altitude aircraft, the difference between the slant range and the map range can be significant at longer distances.

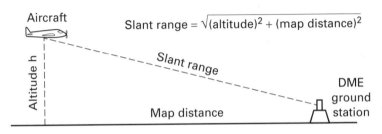

Figure 5–17 Geometry of a DME range measurement.

Example

At what map distance from a DME ground station will an aircraft flying at 35,000 feet (ft) above the ground experience a difference of 25% between slant range and map range?

Solution. First we must recognize that the slant range is always greater than the map range. Thus, the distance at which the error is 25% is where the slant range is 1.25 times the map range. To find this value, write the equation for slant range in terms of altitude and map range.

$$R = \sqrt{a^2 + m^2}$$

$$1.25m = \sqrt{a^2 + m^2} = \sqrt{(5.75)^2 + m^2}, \qquad 0.5625m^2 = (5.75)^2$$

where R is the slant range, a is the altitude, and m is the map range. The units can be anything, meters, feet, or nautical miles, as long as the units are the same for m, a, and R.

In this equation, 5.75 nmi is the equivalent of 35,000 ft. By squaring both sides of the equation and rearranging terms, m is found to be 7.7 nmi, while the slant range is 9.58 nmi. Because of situations such as this, pilots report positions when the distance is measured from a DME as ×× miles, DME.

For an aircraft at 35,000 ft and passing over a DME station, when the aircraft is closer than 9.58 nmi DME, which represents a total of 19.2 nmi DME (9.58 nmi DME before station passage and 9.58 nmi DME after station passage), the difference between the map range and slant range is 25% or greater. At station passage, the slant range read from the DME does not represent the map range at all. The DME will read 5.75 nmi, the altitude, while the map range is zero.

The airborne equipment is called an *interrogator* and transmits pairs of RF pulses at a frequency near 1 gigahertz (GHz). When a ground transponder receives the RF pulses, it checks to be certain that the pulse pairs have the correct time separation for that transponder; if the interrogation is correct, a 50-μs delay time is deliberately introduced, and a reply consisting of a similar pulse pair is transmitted at a frequency different from the interrogation. A different frequency is used so that the airborne receiver will receive only transmissions from the ground station, rather than a mixture of ground and airborne transmissions in which the signals have a wide variation in strength. The ground station receiver does have to contend with a wide variation of signal strengths from the aircraft, but this receiver is a much more sophisticated unit. Each DME installation, both airborne and ground, requires two frequencies: the air to ground frequency, or the interrogation frequency; and the ground to air frequency, or the reply frequency.

Since a number of aircraft are transmitting their pulse pairs to a ground transponder, an airborne interrogator will receive replies intended not only for it but for other aircraft as well. Replies to an airborne interrogator intended for other aircraft and pulses sent by the DME transponder not in response to any interrogation are called *squitter*. The DME interrogator must separate the squitter from the reply. Following an interrogation, a number of pulse transmissions will be received from the transponder; only one of the pulse transmissions is the desired reply.

The elapsed time from an interrogation to a reply from the ground transponder should be roughly the same for subsequent interrogations. The only change in time between interrogations will be due to the motion of the aircraft. Even for aircraft traveling at nearly the

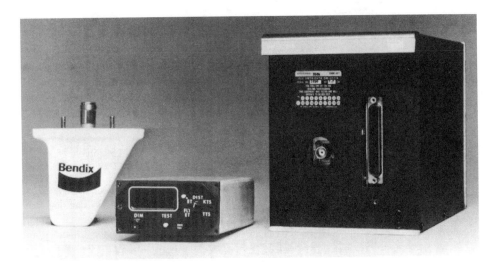

Photo 5–5   A complete DME system showing the receiver/transmitter box, the antenna, and control head. Photo courtesy of Bendix Corporation.

speed of sound, 1000 ft/s, and interrogating the transponder at 16 interrogations per second, the distance the aircraft has moved in the time between interrogations is 63 ft, or 0.01 nmi. Essentially, the elapsed time from interrogation to receipt of the reply is the same for subsequent interrogations.

The method of separating undesired replies is simple. A number of interrogations are made, and the replies resulting from these interrogations are analyzed. The analysis of the replies should show that one reply for each interrogation arrived at roughly the same elapsed time after the interrogation. The remaining received pulse pairs arrive at random delay times and are squitter.

To ensure that the undesired replies are randomly distributed in time and that two aircraft cannot become synchronized, the rate of interrogation from each aircraft must be irregular. Otherwise, if two interrogators had exactly the same interrogation rate, which would be the case if they were the same type of DME, the replies for two aircraft would arrive at the same elapsed time for each interrogation, and the separation technique would not work. DME interrogators have a maximum average interrogation rate of either 30 pulse pairs per second (PP/s) or 16 PP/s. The lower 16 PP/s applies to DMEs TSOed to the later TSO requirements. The TSO requirements were changed because proliferation of DMEs was increasing congestion on the DME frequencies, and modern technology allowed accurate distance measurement to be made with fewer interrogations.

Most DME interrogators operate with a specific interrogation rate, typically the maximum allowed by TSO, or 16 interrogations per second. To have an average interrogation rate of 16 while maintaining an irregular interrogation rate, the time between interrogations is "jittered" from the nominal. The average time between a 16-interrogation per second rate is 62.5 milliseconds (ms). Although the time between each interrogation is different, the average is 62.5 ms. The variation from the nominal is random, being short half of the time and long the other half; thus, on average, the time between interrogations is 62.5 ms.

DME is actually the distance portion of a military navigation system called TACAN, which stands for tactical air navigation. TACAN and VOR may be installed at a common location, and civilian users may use the TACAN system. Most civilian users take advantage of only the DME portion of the TACAN system and use the distance in conjunction with the VOR bearing. When a VOR and TACAN system are colocated, the navigation aid is called a VORTAC. In a few installations, only the DME portion of TACAN is installed and the combination is called a VOR-DME. DME is also installed in many ILS installations.

There are as many DME frequency pairs as there are VOR and ILS frequencies. There are 100 interrogation frequencies and 200 reply frequencies. Since there are 200 VOR and ILS channels, the 100 interrogation frequencies must be shared. This is done by using two different interrogation pulse separations. The channels ending in an exact 100 kHz are called X channels and those channels ending in 50 kHz are Y channels. As an example, 108.00 and 108.10 are X channels, whereas 108.05 MHz and 108.15 MHz are Y channels. Notice that 108.00 and 108.05 MHz are VOR frequencies, whereas 108.10 and 108.15 MHz are localizer frequencies. Although the VOR and localizer signals are quite different, the DME that operates with a VOR has exactly the same signal characteristics as a localizer DME.

The DME interrogation signal envelope appears in Fig. 5-18. To reduce the amount of bandwidth occupied by the DME interrogator and transponder transmitters, the pulses transmitted by both interrogator and transponder are not rectangular pulses with rapid rise and fall times, but are carefully shaped pulses that have a Gaussian shape. Essentially, the Gaussian pulse allows the fastest rise and fall times with the least amount of occupied spectrum.

The pulse separation for an X channel interrogation is 12 μs, and the Y channel interrogation is 36 μs. Because X and Y interrogations share the same channel frequency, the transponder receiver discriminates between these two interrogation pulse separations. DME transponders located at X channel VOR facilities reply only to the 12-μs pulse pairs, and Y channel DME transponders reply to 36-μs pulse pairs.

After the receipt of a valid interrogation, the transponder waits 50 μs, in the case of an X channel, as measured from the half-amplitude point of the first interrogation pulse to

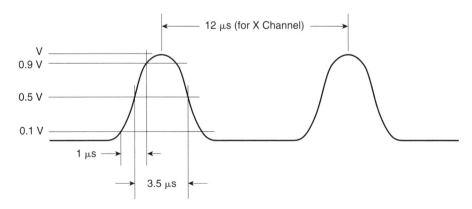

Figure 5–18   DME interrogation and reply pulse shape.

the first reply pulse, and transmits a two-pulse reply. For the X channel transponder, the reply frequency is 63 MHz below the interrogation frequency with a pulse separation of 12 μs, whereas the reply frequency for a Y channel is 63 MHz above the interrogation frequency with a pulse separation of 30 μs. The delay time is 56 μs for the Y channel. The interrogation frequencies span from 1041 to 1150 MHz, and the reply frequencies span from 978 to 1213 MHz. The interrogation frequencies start at 1041 MHz corresponding to 108.00 and 108.05 MHz and occur every 1 MHz. Thus, the interrogation frequency of 1042 MHz corresponds to 108.10 and 108.15 MHz. There is a gap in this frequency range to protect the air traffic control secondary radar system operating at 1030 and 1090 MHz. Therefore, of the 110 MHz between 1041 and 1150 MHz, only 100 channels are used, allowing for 10 unused channels to provide the radar system with the necessary protection. The military TACAN system does have channels in these gaps, but these frequencies are only used for military applications and only rarely.

If the DME transponder transmitted only when replying to interrogations, the number of pulses transmitted by the transponder would be proportional to the number of aircraft interrogating the transponder. A variable rate of transmission from the transponder would cause operational difficulty for the TACAN system, as well as other problems. Therefore, random dummy pulses are transmitted by the transponder when needed so that the same number of pulse pairs, 2700, is transmitted every second. The dummy pulses are replaced with the replies when a reply is due.

The transmitted output from the DME ground station includes random squitter and replies that also occur at random intervals. A regular, meaning not random, set of pulses is transmitted every 37 s; it contains 1350 pairs of pairs per second, which results in the same 2700 pulse pairs per second. Because the pairs of pairs are at a regular rate, if the output of the DME receiver is applied to a loudspeaker or headphone, a 1350-Hz tone will be heard. This regular pulse repetition rate is on/off modulated with Morse code and is used to transmit the DME's station identifier.

At times, a ground transponder will not transmit a reply to a valid interrogation. One such case is when an interrogation is received during the transmission of an ident signal. The interruption of the ident to transmit a reply would result in a distorted Morse code ident. Also, if two interrogations arrived at the transponder at the same time, they would mutually interfere, and a reply could not be generated. Another situation arises when a ground transponder is operating with a very large number of aircraft such that more than 2700 interrogations are received, which would involve only 90 aircraft using the older standard of 30 interrogations per second. Because the ground station transmits an average of 2700 pulse pairs per second, it could not transmit a reply to all the interrogations. In this situation, the transponder eliminates the replies from the weakest interrogations from the most distant interrogators.

Every DME interrogator will be missing some replies because of the reasons outlined in the previous paragraph. The ratio of received replies to transmitted interrogations times 100% is called the *reply efficiency*. DME interrogators can perform to the required specifications with reply efficiencies of 70% or less.

DME signal integrity is assured by the usual ground monitoring stations, as well as by the pilot listening to the ident. The DME also monitors the reply efficiency, determines the signal usability, and flags the DME indicator if the signal is unusable. A number of methods are used to flag an indicator. Typically, the distance display is removed and replaced with dashes, or it is blanked out completely.

DME Hardware

The block diagram of a DME interrogator is shown in Fig. 5-19. A synthesizer provides 100 frequencies for the interrogator transmitter and the local oscillator for the receiver. To share the frequency synthesizer between transmitter and receiver, it is necessary that the receiver IF be equal to the difference in interrogation and reply frequencies which is 63 MHz. Also, because there are two possible reply frequencies for each interrogation frequency, the X and Y channels, the same synthesizer frequency is used to supply two reply frequencies. For an X channel, the local oscillator frequency, or the interrogator frequency minus the IF, or 63 MHz is the X channel, while the Y channel is the interrogation frequency plus the IF.

Example

What is the interrogation frequency and the X and Y reply frequencies for the DME paired with a 108.0-MHz VOR?

Solution.   The interrogation frequency is 1041 MHz. The X channel reply is $1041 - 63 = 978$ MHz, while the Y channel reply frequency is $1041 + 63 = 1104$ MHz. Notice that these two frequencies are images of each other. When the DME is operating on the X channel at 978 MHz, the image frequency is $978 + 2 \times (IF) = 1104$ MHz, which is the Y channel receive frequency.

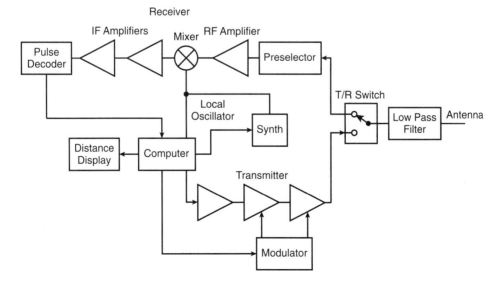

Figure 5–19   Block diagram of a DME interrogator.

The receiver is a single-conversion superheterodyne with a 63-MHz IF. The receiver has a pulse detector and a pulse decoder, which provide an output only if the pulses are of the correct pulse width and separation. The pulse detector also contains a pulse amplitude detector, which is used for controlling the gain of the receiver through an AGC system.

The transmitter amplifies the output of the synthesizer to a level suitable for modulation, and the pulse-modulated output is fed to the antenna. The same antenna is used for both transmission and reception; therefore, an electronic switch is used to transfer the antenna from transmitter to receiver.

A microcomputer is used to measure the time delay, subtract the offset of 50 μs, and convert the result to miles for the display or interface to other navigation equipment aboard the aircraft.

The DME also provides a ground speed by taking the change of distance divided by the time between interrogations.

Most DMEs have a memory that will hold the distance if, for a number of reasons, the DME has received no valid replies. This lack of valid replies does not apply to a short-term situation, which occurs regularly and results in a reply efficiency of less than 100%, but to a long-term lack of replies. This lack of replies can occur when the aircraft is banked such that the DME ground station is blocked from the airborne antenna. It can also occur when the Morse code ident is transmitted. During the ident transmission, the DME transponder does not reply to interrogations. To provide a continuous display of distance and to prevent the DME from having to acquire lock after a short loss of signal, the distance is stored in a memory and is displayed for up to 10 s after a loss of signal.

LONG-RANGE NAVIGATION

A form of radio navigation that provides continuous navigation information over much of the world is LORAN-C. LORAN-C is an improved version of an old long-range navigation system, LORAN. LORAN stands for LOng RAnge Navigation and was originally installed in the 1940s for use by ships at sea. All LORAN systems operate with the transmission of pulses from several time-synchronized transmitters.

The LORAN-C system has a master transmitter and several slave stations. The combination of a master and its slaves is called a *chain*. The master transmits a series of pulses at a frequency of 100 kHz and, after a delay time, the first slave transmits a similar series of pulses at the same frequency. After an additional delay time, a second slave transmits a series of pulses. This sequence continues until all the slaves have transmitted. The time delays are arranged so that the master is always received first, followed by slave 1, slave 2, and so on.

The earlier LORAN systems required the slaves to receive the pulse from the master before it transmitted its pulse. This introduced various errors because of the small variations in propagation between master and slaves. Modern LORAN-C systems use atomic clocks to time masters and slaves, which removes the variable propagation as an error term. Monitoring stations are used to detect any drift of the atomic clocks and to provide correction.

As each transmission is made, the transmitted energy moves away from the transmitter in ever-widening circles, as shown in Fig. 5-20. Since the slaves do not transmit until a fixed time delay after the master transmits, the circles emanating from the slave are delayed. By measuring the time delay from a master and a slave, the position of the LORAN-C receiver can be determined to a line. To understand this, consider the 4-ms circle from the master in Fig. 5-20 and the 5-ms circle from the first slave. This represents a 1-ms time differential. However, the intersection of the 5-ms circle from the master and the 6-ms from the first slave, and of the 6 and 7, and of the 7 and 8, and so on, all represent a 1-ms time delay. The figure shows circles at even millisecond intervals. Of course, we could draw an infinite number of circles representing time to any level of accuracy desired. This implies that there are an infinite number of points at which the time difference is the same between the master and the slave. All the points that represent a 1-ms time differential fall on a hyperbole. Measuring the time delays between the master and one slave locates the receiver to a line of position, or LOP.

A second line of position may be determined by using the time delay between the master and a second slave. Two hyperboles can intersect at two points, and a third LOP is required to locate the receiver to a unique point. This third LOP can be determined from a third slave or from the time delay between the two slaves. Because the transmissions of the master and all slaves of a chain are absolutely synchronized, any two members of a chain will produce an LOP. However, in many cases, if the approximate position is known, such

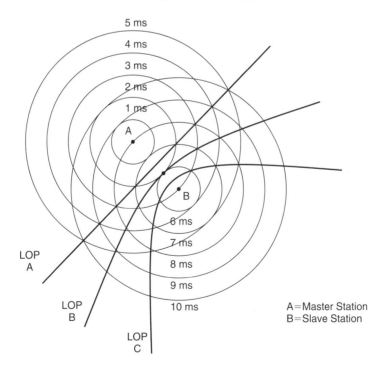

Figure 5–20   LORAN-C propagation characteristics.

as when the receiver is tracking a master and three slaves and one slave is temporarily lost because of interference or a weak signal, the second intersection point is so far removed from the actual position that it can be ignored.

When the receiver is turned on, there may not be any information relative to the position and three LOPs will have to be determined to make the initial fix.

Every LORAN-C chain operates at one frequency, 100 kHz. To identify the LORAN-C chains, each chain has a unique repetition rate, which is distinguished by the time between the transmission of the master and slaves, called the *group repetition interval,* or GRI. The shortest GRI is 49.9 ms, which represents a transmission rate of about 20 Hz; the longest GRI is 99.9 ms, which represents a transmission rate of about 10 Hz. The standard representation for a GRI is to multiply the time in milliseconds by 100. As an example, the shortest and longest GRIs are represented by the 4990 and the 9990 chains, respectively. Because the GRIs are different, any interference that occurs because of the reception of a signal from more than one chain will not occur on the next transmission because the two chains have different intervals between transmissions.

Notice that the GRIs are all multiples of 100 μs, which is the time of exactly 10 LORAN-C carrier cycles. This implies that a LORAN-C transmission always begins with the zero crossing of the carrier. The entire LORAN-C system is synchronized so that the GRIs and carrier cycles are all phase coherent and are under the control of the precise atomic clocks. This permits both the envelope and carrier phase of the LORAN-C transmissions to be used for time delay measurements.

Example

Typically, a master is more than 500 nmi from any of its slaves. What minimum time delay is required from master to slave to ensure that the slave is received after the master anywhere in the coverage area of the LORAN-C chain?

Solution.   If the RF pulse from the master has arrived at the slave location, the slave may transmit any time after that and thus be sure to arrive after the master pulse. Thus, it is necessary to calculate the time of propagation of a radio wave from the master to the slave or a distance of 500 nmi.

The distance of 500 nmi is 1853×500 or 9.27×10^5 m. The velocity of radio waves is 3×10^8 m/s. Dividing the distance in meters by the velocity of radio energy produces the result:

$$\frac{9.27 \times 10^5}{3 \times 10^8} = 3.09 \text{ ms}$$

Therefore, a delay of 3.09 ms minimum must be inserted from the transmission of the master pulse to the transmission of the slave pulse.

To further identify the master, the master transmits a group of eight pulses separated by 1 ms and a ninth pulse 2 ms after the pulse. The slave stations, on the other hand, only transmit eight pulses, as shown in Fig. 5-21.

The LORAN-C pulse is shown in Fig. 5-22. The pulse is carefully shaped to provide the maximum rise time that can be used for the time measurements without causing excessive radio-frequency spectrum. This is similar to what is done with the DME pulses except

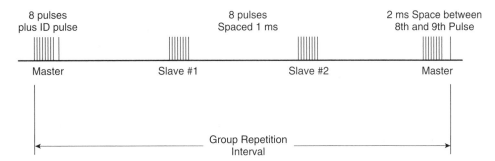

Figure 5–21 Master and slave pulses of a LORAN-C chain.

that, in the case of the LORAN-C system, the actual carrier zero crossings are used, whereas the high, 1000-MHz frequency used for DME precludes this.

At the airborne receiver, the received LORAN-C pulse may have propagated to the receiver by one of two paths. The first, and the desired, path is by ground wave. However, some of the received energy may have arrived by sky wave, which is an undesired mode of propagation. When the sky wave arrives, it contaminates the ground wave and causes a distorted waveform.

Fortunately, the sky wave does not arrive sooner than 30 μs after the ground wave, which allows the first three cycles of the LORAN-C waveform to be received undistorted. The third cycle is the greatest amplitude of the three and is used for timing.

LORAN-C Hardware

The LORAN-C receiver is a single-frequency receiver with no demodulator. The LORAN-C receiver is, essentially, an amplifier with some selectivity to pass the LORAN-C pulses without distorting the received signal. Although the LORAN-C receiver looks something like a

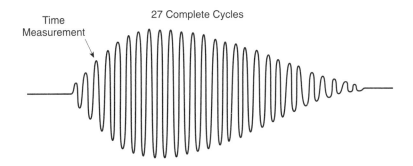

Figure 5–22 LORAN-C pulse.

TRF, it has no demodulator. The received pulses are not demodulated, but the zero crossings of the actual carrier are used for timing.

Figure 5-23 shows the block diagram of a LORAN-C receiver. The actual timing
is done relative to the zero crossing of the third cycle of the LORAN-C pulse. The third
cycle of the pulse is determined by an envelope detector, and the zero crossing initiates the
timing.

Because the LORAN-C signal is a frequency sufficiently low that conventional A/D
converts may be employed to digitize the signal, it is common for a LORAN-C receiver to
digitize the actual received signal and perform the signal processing using a microcomputer.

The LORAN-C receiver employs a computer to convert the time delay between the
master and the received slaves to a point of position. In this process, the computer first must
identify the chain being received, look up the stored data relative to that chain, and determine the latitude and longitude of the master and the slaves. The point of position in latitude and longitude is calculated relative to the latitude and longitude of the master and
slaves.

There is a very important difference between the LORAN-C system and the radio navigation systems discussed so far, the NDB/ADF, TACAN, and VOR/DME. The previously
discussed systems provided navigation information relative to a point on the ground, the
VOR station or the NDB. These ground navigation aids are located at strategic locations
such as on or near victor airways or near airports. The LORAN-C system provides latitude
and longitude.

To an aviator, latitude and longitude are of little value. Thus, the LORAN-C receiver
accepts position information for *waypoints* as latitude and longitude and then provides the

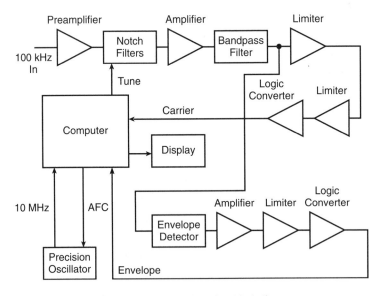

Figure 5-23   LORAN-C receiver block diagram.

required heading, steering information, and distance to the waypoint much like a VOR and DME. Other information, such as the the ground speed and track errors can be calculated. Unlike the VOR, a waypoint can be created to lie anywhere, and the aircraft may be flown as if the aircraft were flying to a VOR.

The LORAN-C system is capable of providing an accuracy of about 500 ft with good-quality signals. The repeatability of the LORAN-C system is typically 50 ft. Thus if an airport is approached for the first time, the accuracy of the LORAN-C fix will be 500 ft. If the latitude and longitude as determined by the LORAN-C receiver are stored as a waypoint, and most LORAN-C receivers have literally hundreds of waypoints for this purpose, the airport may be approached for future trips to an accuracy of 50 ft.

GLOBAL NAVIGATION

Although LORAN-C provides a very large coverage area, in some areas in the world there is no LORAN-C coverage. Another long-range, radio-based navigation system called *Omega* provides full global coverage.

Omega is similar to LORAN in that it is a hyperbolic navigation system. One difference is that the frequencies used for Omega are much lower than those used for LORAN and, consequently, the wavelength is much longer. The wavelengths are so great that they just about fit between Earth's surface and the conducting layers of the ionosphere. Therefore, the waves are constrained in this region like water trapped in a pipe. Because of this, there is no sky wave to cause distortion, and the energy lost is very low. There is also much less atmospheric noise in this frequency range. Only eight transmitters operating with much lower powers than LORAN-C provide full global coverage.

The Omega system relies on continuous-wave transmission from each of the eight transmitters. Four common transmitting frequencies are used by each station and one unique frequency is used by each station. Each station transmits on a unique sequence of frequencies that is recognized to identify the station.

Considering the entire transmission cycle, as shown in Fig. 5-24, during eight time periods all four of the common frequencies are being transmitted by four of the eight stations. The four common operating frequencies are all harmonics of a common 283.333-Hz frequency. Thus, every 1/283.333 s, 60/17 ms, all the Omega transmitter carriers together cross zero on a rising edge. This reference point is used for making phase measurements. This phase measurement is used similarly to the time measurement for LORAN-C.

There is a difference between a phase measurement and a time measurement. The phase measurement repeats every wavelength. Therefore, the same phase measurement will exist one wavelength away. For the wavelengths associated with Omega, this repeat measurement occurs in as short a distance as 12 nmi, which corresponds to the wavelength of the highest Omega transmitting frequency. Measuring the phase angle between two Omega transmitters as received by the airborne receiver will determine position, like the LORAN-C, to a line of position, LOP. However, there are a number of LOPs because the phase measurement repeats every wavelength. By taking a second set of phase measurements from the same two Omega transmitters, but at another of the four operating frequencies, a sec-

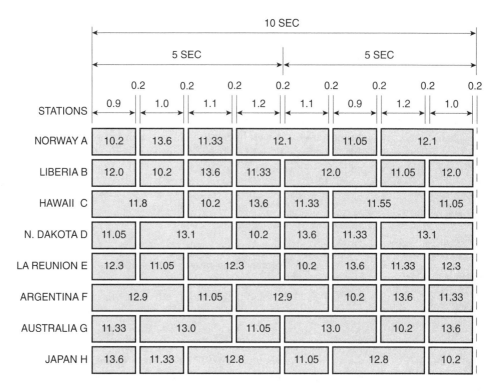

Figure 5–24   Transmission sequence for the Omega system.

ond set of LOPs can be determined that repeat each wavelength. Because this represents a different wavelength, this second set of LOPs does not correspond to the first set of LOPs. The "real" LOP is where the LOPs determined by both phase measurements coincide. If two sets of LOPs are determined relative to 13.6 and 11.333 kHz, every fifth wavelength of 13.6 kHz will correspond with every sixth wavelength of 11.333 kHz. Therefore, there is a real LOP every 60 nmi.

If a third set of LOPs is measured using a third frequency, say 10.2 kHz, LOPs corresponding to 12 wavelengths of 13.6 kHz, 10 wavelengths of 11.33 kHz, and 9 wavelengths of 10.2 kHz coincide. The ambiguity has been reduced to 144 nmi. With the use of the fourth frequency, 11.05 kHz, the ambiguity is increased to 571 nmi, or, effectively, no ambiguity.

Usually, an Omega receiver remembers its last position fix and can easily remove any ambiguities. When an Omega system is turned on for the first time, a situation called a *cold start,* it may be necessary to enter the aircraft's position to eliminate any ambiguities.

Figure 5-25 is the block diagram of an Omega receiver. The receiver sequentially receives each frequency and measures the phase between the received signal and the receiver's internal clock. The receiver's internal phase standard has no specific phase reference; it is just used as a local reference to measure the phase angle between the received signals. Because there are eight Omega stations and each transmits on the four common frequencies, there can be as many as 32 phase measurements. Usually not all eight stations provide reliable signals, and therefore not all 32 measurements are made. The phase measurements that are taken are used in various combinations to reduce the ambiguity and to find a number of LOPs to determine the aircraft's position.

Because in the Omega system some received signals have traveled a very long distance, any variation in the velocity of propagation will result in an error of position. Many Omega receivers have the capability to make a number of corrections based on the amount of land or seawater the signals have traveled over, the state of the ionosphere that changes with the 11-year sunspot cycle, and the amount of signal that has passed through daylight or darkness. To enter these corrections, the Omega receiver must know the date, the time of day, and an approximate position. An internal clock/calendar provides these data in order for the receiver to make the corrections.

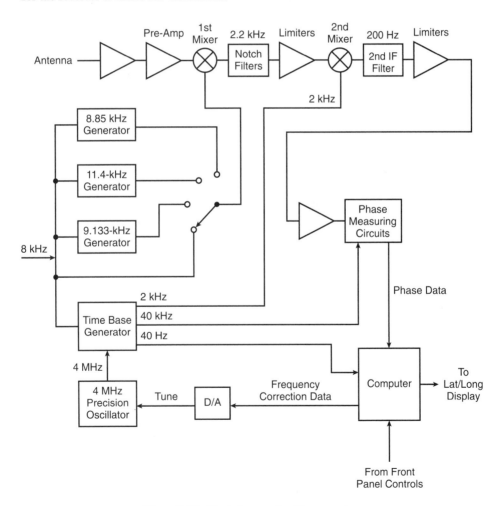

Figure 5–25 Block diagram of an Omega receiver.

Photo 5–6   The Ashtech GPS receiver provides a moving map display for navigation. Photo courtesy of Ashtech, Inc.

As for the LORAN-C receiver, the latitude and longitude of the present position are converted to steering information similar to a VOR indicator.

SATELLITE NAVIGATION

The most advanced form of radio navigation is satellite navigation. Two systems are in operation at the time of writing: the U.S. system, Global Positioning System, or GPS, and the Russian system, Global Orbiting Navigation Satellite System, or GLONASS.

Both satellite navigation systems are elapsed-time multiple ranging schemes. Unlike earthbound navigation systems, the satellites are in constant motion, and the radio signal propagation is absolutely line of sight, rather than following Earth's curvature. Because of the direct propagation, the satellite navigation systems can provide a three-dimensional position determination anywhere on Earth. This text will discuss the U.S. system.

The GPS system consists of three elements, the spaceborne segment, the ground-based monitoring stations, and the navigation receivers, or the user segment.

The spaceborne segment consists of a total of 24 satellites; 21 are used for navigation and 3 are reserved for spares. The satellites orbit in six orbital planes at a distance of 10,900 nmi above Earth's surface, which results in exactly two orbits each sidereal day. A sidereal day is the time Earth takes to make one revolution. The normal concept of a day is from high noon to high noon. The motion of the sun in the sky is a combination of the rotation of Earth on its axis and the effect of the motion of the Earth around the sun. The time it takes for Earth to make one revolution on its axis is 23 hours and 56 minutes.

Because the orbits of the satellites are synchronized with the rotation of Earth, the ground track of the satellites repeat. Therefore, an equation can be used to predict the position of the satellite relative to the ground if the precise time is known.

The satellites send continuous data including the time of day, according to the satellite's clock, when the satellite transmission was made. The satellite has a very accurate atomic time-of-day clock.

The ground segment of the system receives the transmitted data from the satellites and monitors the position of the satellites in their orbits. A number of factors can disrupt the orbits of the satellite such that they deviate from the calculated position. The constants for the equations used to locate the satellite in its orbit are relayed daily to the satellite. Because of perturbed orbits, the daily changing of the constants allows for the equations to accurately predict the position of the satellites. These constants, called *ephemerides,* are included in the data message sent to the user segment.

The GPS receiver is the user equipment installed in the aircraft. The GPS user equipment receives the transmissions from the satellites and decodes the identity of the satellite, the time of day, and other data concerning the satellite. The ephemerides and the time of day when the transmission left the satellite are used to calculate the position of the satellite.

The time of day of the transmission is subtracted from the GPS receiver's clock, and the elapsed time from the transmission of the GPS message to the receipt of the transmission by the GPS receiver is calculated. This elapsed time is divided by the speed of light to provide the distance between the satellite and the user vehicle. This distance is called *pseudorange.* The prefix *pseudo-* is used because corrections must be made before the range is useful for navigation. Thus, by receiving the transmission from one satellite, the position of the GPS receiver can be placed on a sphere, called a *sphere of position,* with the satellite at the center, as shown in Fig. 5-26.

By receiving a second satellite, a second pseudorange is measured, and the position of the receiver can be placed on a second sphere of position, as shown in Fig. 5-27. The two spheres of position intersect, which produces a circle that reduces the ambiguity from a sphere to a circle.

By receiving a third satellite, a third sphere of position is determined, as shown in Fig. 5-28, that intersects the circle at one point. This point was determined from pseudorange measurements and does not represent a useful navigation fix without correction.

The major source of error in the range measurement, causing it to be called pseudorange, is the GPS receiver's time of day clock. Because the GPS receiver does not contain an atomic clock, the time of day error can easily be a significant fraction of 1 s, depending

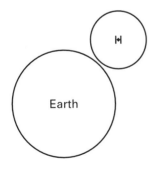

Figure 5–26   Sphere of position as determined from the range measured from one satellite.

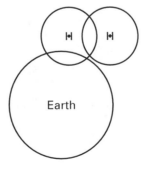

Figure 5–27   Circle of position determined from two satellites.

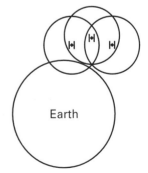

Figure 5–28   Point of position as found with three satellites.

on how long it has been since the GPS receiver was last used. How can the user clock be set if the GPS receiver has been off for some time? First, the nominal distance to a satellite is 10,900 nmi, which requires 67 ms to propagate a signal from the satellite to the user segment. The satellites received are at different distances from the user and thus experience different elapsed times. A simple first approximation at setting the user clock is to take the average of all the times received from the satellite and assume an average propagation delay of 67 ms.

To better understand how much the clock accuracy can affect the range measurement, consider the fact that radio waves travel about one-third of a meter in 1 nanosecond (ns). Therefore, if the clock error is, say, one-thousanth of a second, the error of the range measurement is 3×10^5 meters, which is a range error of about 162 nmi. Because the range to the satellite is about 10,900 nmi, the error is rather small. However, navigation accuracy of 162 nmi is not acceptable and this error must be corrected.

To correct the GPS receiver's local clock to determine a point of position that is accurate, a fourth satellite is received and a pseudorange is determined from the fourth satellite. This pseudorange can be used with the other three to determine three more points of position. If the satellites are A, B, C, and D, the four points can be determined from the four possible combinations of three satellites, ABC, ABD, ACD, and BCD. If there are no errors from any source, the four points of position will coincide. A lack of coincidence can be caused by any error source, most notably the GPS receiver clock. Corrections are made to the clock and the points are checked for coincidence. When the spread of the points of posi-

Photo 5–7   The Universal navigation management system includes a navigation computer, a GPS receiver and control and display functions. An 80,000 waypoint memory is loaded from the front panel using a plug-in update card. Photo courtesy of Universal Navigation Corporation.

tion has been reduced to the minimum, the clock has been corrected. By using this technique, the clock error can be reduced to a few nanoseconds. The range calculations made from the corrected clock are no longer pseudorange but may be used for solving the position equation.

The use of four satellites is required to correct the GPS receiver's clock, and the use of more satellites can improve the accuracy of the system by allowing data averaging. A minimum of six satellites is visible and often as many as eight are available. It is usually not beneficial to use all the visible satellites, because some of the satellites, such as those very low on the horizon, will have considerable error because of the bad geometry. Adding satellites with bad geometry will degrade the accuracy rather than enhance it. This phenomenon is called *geometrical dilution of precision* or GDOP. Many GPS receivers will receive and track all visible satellites but will only use those satellites that can improve the navigation precision.

The satellites transmit on one common frequency. If each satellite is assigned to a specific frequency, 24 frequencies will be required. Of these frequencies, more than half will go unused at any one time because more than half of the satellites are not visible.

Another technique for providing communications for all 24 satellites is to use a technique called *spread spectrum*. This is the opposite of assigning each satellite to a specific frequency. Spread spectrum causes all 24 satellites to operate on the same band of frequencies, and each satellite uses the entire band of frequencies.

The data transmitted by the satellite are only 50 bits per second, which requires only a narrow frequency band. The satellite signal is deliberately spread to a 2.046-MHz frequency range by modulating the narrowband data signal with a noiselike signal. Because the bandwidth of the noise is on the order of 1.023 MHz, the net result is a signal with a bandwidth of 2.046 MHz centered on 1.575 GHz, because both upper and lower sidebands are generated.

The noiselike signal is not noise, but follows a predictable pattern if the generating algorithm is known. The GPS signal appears to be random, but actually repeats every 1023 bits or 1 ms. The sequence is called the pseudonoise code or PN code.

Because the receiver knows the generating algorithm, it can remove the noise modulation from the wideband signal and retrieve the modulation. The receiver "unmodulates" the PN code from the transmission and recovers the original data modulation. This technique is called *despreading.*

All the GPS satellites share the same band of frequencies, but each satellite has its own PN sequence. During the correlation process, only the data from the satellite matching the PN code used for correlation will be extracted. The other satellites will appear as any other noise source.

Not only does the recovering process require the exact same PN sequence, but it must be accurately time synchronized. Since the PN bits change every 1 μs, the local PN code must be aligned to within 1 bit of 1 μs to despread the signal. This time synchronization provides additional time measurement accuracy. Therefore, the despreading not only recovers the encoded data, but provides precision timing to determine the pseudorange. The uncovered data plus the timing information due to the despreading process are used to calculate the point of position as latitude and longitude.

As for the LORAN-C and Omega receivers, the latitude and longitude of the aircraft are not useful. Steering information is provided relative to a waypoint that is entered into the GPS receiver.

Two levels of accuracy are available from the satellite system. Using the signals transmitted at 1.575 GHz, the *standard precision service,* on SPS, is obtained. There is a more precise mode of operation called the *precision position service* or PPS. The service involves a second carrier frequency at 1.22 GHz and a second PN code. This service is not available to civilian users.

The GPS system was constructed by the U.S. Department of Defense (DoD) and is primarily a military system. To prevent hostile forces from using the GPS system for action against the United States, the DoD has methods of making the satellite system unusable to unauthorized users.

The first prevention is simply to not disclose the PN codes required for the PPS mode. However, even the performance of the SPS is impressive and could be used by hostile forces. To prevent this, the satellites can be deliberately degraded by a technique called *selective availability,* or SA.

SA causes the satellites to tell lies. The most effective use of SA causes the time of day transmitted by the satellite to be incorrect. The deliberate error is random and constantly changing, so a receiver cannot discover the error. This is called *clock dithering* and the amount of error introduced by this technique can be adjusted. Remember that the satellites are under complete control of the ground segment, and SA can be turned off or on and adjusted to whatever level of precision or, more accurately, nonprecision is desired.

GPS Precision

When GPS was first introduced, the phenomenal accuracies reported impressed everyone. However there are two errors of position, horizontal and vertical. When SA is operating, the error is a function of the amount of SA applied. Typically, the horizontal error is 100 m, which is not sufficient for precision approaches but will serve for en route navigation and nonprecision approaches. The vertical accuracy is actually worse at 150 m and makes this part of the GPS position fix useless for any aviation purpose. If SA is turned off, the precision improves dramatically to 30 m horizontal and 45 m vertical. For instrument landings, this is still not sufficient for anything better than the least restrictive of precision approaches.

There are other error sources that cannot be turned off. First, the speed of light in the atmosphere is a variable and cannot be measured by receiving the SPS transmission only. The PPS mode uses two carriers at two frequencies. Having the two frequencies allows variations in propagation velocity to be measured and used to correct the range measurements for enhanced accuracy. The PPS also uses a higher-speed PN code that is inherently more accurate. The phenomenal accuracies reported for GPS, in some cases measured in centimeters, are for surveying applications. These precisions were obtained with a very long period of data averaging, sometimes as long as an hour. In some cases, the extreme precision was for the PPS, which is not available to civilian users. Also, carrier phase tracking, which measures the number of wavelengths of the transmitted carrier between two antennas, was used for some high-accuracy measurements. Finally, high-accuracy measurements seldom involve moving vehicles.

There is a method of accuracy enhancement that can dramatically improve the basic accuracy of the SPS even with a moving platform and with SA operating. This technique is called *differential GPS* or DGPS.

DGPS uses a known point where a GPS receiver is installed. The GPS ranges are measured from each satellite in view, and the range errors are calculated by taking the difference between the known range and the range measured by the satellite transmission. The range error for each satellite is broadcast to GPS users that can use these corrections to remove the effects of SA and the variation of propagation velocity. The range corrections are continually broadcast, so the variations deliberately introduced by SA are continually removed. The facility that determines the range corrections and broadcasts the correction is called the *reference locator.*

Using DGPS, accuracies approaching that available with PPS are possible. An accuracy of better than 3 m horizontal and 5 m vertical may be achieved using DGPS. This opens the possibility of using DGPS for precision landings.

At the time of writing, there is no national standard for a reference locator for aircraft use. A national standard would specify such characteristics as the carrier frequency of the correction broadcasts, the protocols of the data, and the modulation method. The U.S. Coast Guard has provided DGPS service for shipboard use and has established standards, but these reference locators are not suited for airborne use.

NAVIGATION COMPUTERS

When an aircraft is equipped with only a VOR navigation system, the only navigation information is course deviation and the selected radial. The aircraft flies from VOR to VOR and, because of this, standard airways or victor airways stretched from VOR to VOR.

However, there are situations where a desired course may not lie on a victor airway, and a number of navigation techniques are used to fly courses off the victor airways. A type of navigation computer was introduced a number of years ago that essentially permits a VOR to be electronically "moved" to any convenient point and the computer calculates the necessary steering and DME information to fly to this relocated VOR. Any number of relocated VORs, called *waypoints,* can be programmed into the navigation computer, and practically any course can be flown.

There is a requirement that a VORTAC or VOR/DME facility be available somewhere in the area to provide the raw information to generate the navigation guidance to the waypoint. This computer is called an *area navigation computer,* or RNAV.

The LORAN-C receiver has a computer to provide VOR-like navigation guidance to a waypoint, but the input data to generate the navigation guidance is from the LORAN-C system. Why can't the LORAN-C and VOR area navigation computer be combined?

The Omega receiver and the GPS also have computers to provide similar navigation information. The obvious arrangement was to provide a computer that uses VOR/DME, Omega, LORAN-C, and GPS information to provide navigation guidance. The computer

Photo 5–8 A hand-held CPS receiver with a detachable antenna for improved reception within an aircraft. Photo courtesy of Magellan Systems Corporation.

analyzes the data, determines which navigation systems have the best geometry and the best signal to noise, and which are, generally, providing the best information. The computer uses the best information and improves, where possible, the accuracy of the navigation solution. If the quality of a signal degrades or the geometry becomes poor, another navigation aid is used to provide the main navigation problem solution.

When a computer provides the navigation guidance, the GPS, LORAN-C, Omega, and VOR/DME systems are called sensors. The actual role of the sensors depends on a number of factors. One factor is signal quality; a poor-quality VOR signal, as an example, will be set aside until the quality improves, or an alternative VOR is chosen while navigation continues using other sensors.

The area navigation computer is an excellent vehicle for storing database information such as VOR locations, VOR frequencies, airport information such as latitude and longitude, standard trips, and waypoints.

Photo 5–9 An integrated navigation system which contains a navigation receiver, a DME interrogator, a complete ILS system, and an RNAV. Photograph courtesy of King Radio Corp.

REVIEW QUESTIONS

5.1. What were the problems associated with the A–N range?

5.2. What has made the NDB an attractive navigation aid for more than 50 years? Where would NDBs most likely be found?

5.3. What are the navigation problems associated with the NDB? How does the VOR system overcome these problems?

5.4. How does the NDB assure that the radio signals transmitted by the station travel by ground wave?

5.5. How is the ambiguity of the peak or null of the loop antenna resolved?

5.6. Through what instrument(s) does the VOR system provide navigation guidance?

5.7. What device of the airborne VOR equipment is used to indicate that the VOR signal is trustworthy and may be used for navigation?

5.8. If an aircraft is heading 45° and flying a VOR radial of 225° TO, is the aircraft flying toward or away from the VOR?

5.9. A DME interrogation is transmitted and a reply arrives 125 μs later as measured from the rising edge of the first interrogation pulse to the rising edge of the first reply pulse. The VOR channel with which the DME is paired is 110.0 MHz. What is the distance to the DME station?

5.10. Relative to DME, what is squitter? Why is it used?

5.11. How is the DME station's Morse code identifier added to the ground station transmission?

5.12. What operating frequency does LORAN-C use?

5.13. Why is the timing for a LORAN-C receiver made from the third cycle of the pulse?

5.14. How are LORAN-C chains identified?

5.15. What situations would make Omega a choice over VOR or LORAN-C for navigation?

5.16. How many Omega transmitters are there?

5.17. At what altitude are GPS satellites? How long is a satellite orbit?

5.18. What ensures that a GPS satellite repeats its ground path each orbit?

5.19. What is selective availability? Why is it used?

5.20. How does differential GPS improve the accuracy of GPS fixes?

5.21. What is GDOP?

5.22. Why are some satellites tracked but not used for the navigation calculations?

5.23. What does an RNAV do?

5.24. What is a sensor as applied to an RNAV?

6

Landing Aids

CHAPTER OBJECTIVES

In this chapter the fundamentals of the localizer and the glide slope will be studied. The student will gain an understanding of all the components of the ILS upon completion of this chapter. The microwave landing system will be understood from an operational viewpoint.

INTRODUCTION

Aviation pioneers involved with early instrument flight knew the importance of landing guidance. Their radio navigation system, although used for all phases of the first instrument flight, was really intended to be a landing aid. These early researchers knew that the most critical part of a flight is the landing, and if a radio navigation aid could be developed that satisfied that portion of the flight, certainly radio navigation could be developed for en route navigation.

For reduced-visibility landing, a system of radio guidance was developed that provides horizontal guidance along the runway center line, called the localizer, vertical guidance for the normal glide path called the glide slope, and radio position marker beacons. In some cases, DME is provided as an adjunct to the localizer. The entire system is called the *instrument landing system,* or ILS.

A more modern system is available at some airports called the *microwave landing system,* or MLS. This system allows making landing approaches more complex than the straight-in approaches permitted using the ILS.

The *localizer* provides two radio signals; each is transmitted from a unique antenna pattern such that the two signals overlap along the runway center line, as shown in Fig. 6-1. The two transmitted signals are identified by two different forms of modulation. The modulation applied to the signal that covers the region to the left of the runway as viewed from the approach is amplitude modulated with a 90-hertz (Hz) sine wave. The signal provided to the right of the runway as viewed from the approach is modulated with a 150-Hz sine wave.

The carriers supplied to the left and right antenna patterns are exactly the same; that is, the carriers are precisely the same frequency and phase coherent. Therefore, there is no interference between carriers because there is only one. Although the carriers are the same, there are two different modulations and thus there are two distinct signals.

To the navigation receiver, the two separate signals combine to become one signal with two modulations, 90 and 150 Hz. The amount of 90- and 150-Hz modulation is a function of where in the antenna pattern the aircraft is located. Clearly, at the center of the antenna overlap, that is, on the runway center line, the 90- and 150-Hz modulation percentages are exactly the same. As the aircraft deviates from the center line, either 90- or 150-Hz modulation will predominate. An expression that quantifies the amount of predomination of modulation is called the *differential depth of modulation,* or DDM, which is given by

$$\text{DDM} = \frac{P_1 - P_2}{100}$$

where P_1 is the greater of the modulation percentages and P_2 is the lesser. Since the DDM does not indicate which of the two modulation frequencies predominates, this must be specified. As an example, a DDM may be 0.4, 150 Hz.

Localizer carrier frequencies are the odd 100 kilohertz (kHz) and odd 100 kHz plus 50-kHz frequencies for the lowest 2 megahertz (MHz) of the navigation frequency band. Thus, valid localizer frequencies are 108.10, 108.15, 108.30, and 108.35 MHz, and so on, which

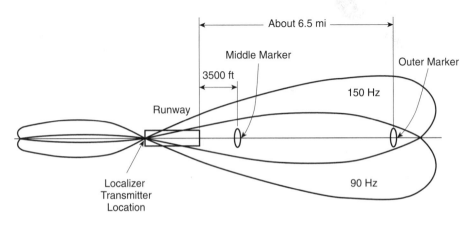

Figure 6–1   Antenna patterns associated with the localizer.

Photo 6–1 The Bendix CN-2011A is a complete ILS system as well as a dual communications and navigation system. Photo courtesy of Bendix Corporation.

results in 40 localizer frequencies. The frequencies of 108.10 and 108.15 MHz are reserved for test purposes and rarely are active navigation facilities assigned to these frequencies.

The localizer Morse code ident is applied to the carrier using 20% modulation and a 1020-Hz tone. The localizer idents have four letters beginning with I, for ILS.

The localizer is usually received with the navigation receiver, the same receiver that is used for receiving VOR. Only in very large aircraft would a receiver be dedicated to localizer reception only. Most navigation receivers cover the range from 108.00 to 117.950 MHz, which covers all VOR and localizer frequencies.

The navigation receiver, which was described in Chapter 4, is used to receive the localizer, and the audio output is fed to a localizer converter, rather than to a VOR signal converter. The output of the localizer converter drives the CDI meter, the same meter used by the VOR. The receiver automatically switches to the localizer converter when a localizer frequency is selected.

Converting the received modulations to an indication of the course deviation indicator can be achieved in a number of ways. The following description is one of the more common methods. The ILS audio is band pass-filtered using two filters, one tuned to 90 Hz and the other to 150 Hz, as shown in Fig. 6-2. The output of each band-pass filter is rectified, one with a positive polarity and the other with a negative polarity. The two outputs are filtered to produce a steady dc voltage.

The localizer course deviation indicator should point in the direction required to return the aircraft to the approach path. Therefore, if the aircraft is to the left of the runway center line, where it receives a predominantly 90-Hz modulation, the indicator should point to

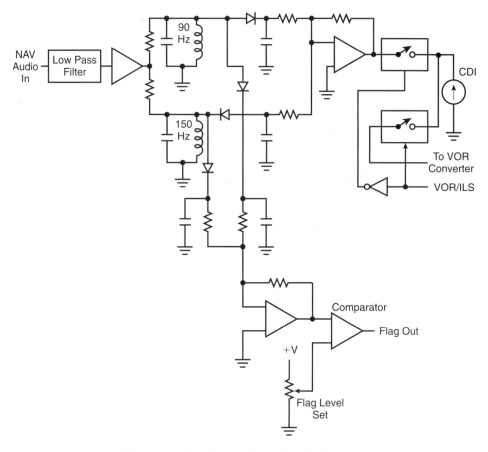

Figure 6–2   Block diagram of a localizer signal converter.

the right because that is the direction required to return to the approach path. Likewise, when to the right of the approach path, the predominant modulation is 150 Hz and the meter points to the left. This is consistent with the meter sense for the VOR.

The output of the 150-Hz rectifier is a positive voltage, whereas the output of the 90-Hz rectifier is a negative voltage. These two voltages are added together, and the result is used to drive the course deviation meter. When the amount of 90- and 150-Hz modulation is the same, or a 0 DDM, the positive and negative voltages from the two rectifiers are the same and produce a net voltage of zero. As soon as the 90-Hz modulation predominates, there is less output from the 150-Hz filter and more at the 90-Hz filter, and thus the net result after the summation is negative voltage, which drives the meter to the right. In the same fashion, when more 150-Hz modulation is present, the meter will swing to the left.

Like any other radio navigation aid, a flag must be provided for the localizer to indicate that the signal is untrustworthy. The flag requires that one or both of the modulating frequencies be present if the flag is pulled from view. The flag circuit is generated very simply by rectifying the two tones with the same polarity and summing the results. The flag prevents the indicator from showing a valid indication when either the modulation level or the signal level is insufficient.

Because the localizer operates with directional antennas, which typically have some radiation from the rear of the antenna, the localizer antenna provides a similar signal to the departure end of the runway. This coverage is called *back course;* it can be used for guidance during departure or for making an approach to the opposite end of the runway.

When using the back course for departure guidance, the left/right signal polarity is the same as for an approach. However, when using the back course for an approach to the opposite end of the runway, four very important points must be taken into consideration. First, the precision available from the back course is less than from a typical ILS approach course. Second, the 90- and 150-Hz signals are reversed. The latter problem can be dealt with by either reversing the meter polarity or simply flying the meter in reverse. The lack of precision must be realized and accounted for. Third, there is no back course equivalent to the glide slope. Finally, there seldom are position markers on the back course.

The directional antenna system used for the localizer also radiates some energy from the side. Because energy can be radiated from the side of both the 90- and 150-Hz antennas, it is possible to receive a weak but convincing false approach path when the aircraft is relatively close to the airport and off to the side of the antenna. This path will lead to the runway but not at the runway heading. To prevent this from occurring, some localizer installations have a set of antennas called *clearance arrays.* These antennas are provided with a high level of 90- and 150-Hz modulation, with the 150-Hz modulation to the right of the runway as viewed from the runway heading and the 90-Hz modulation to the left. These signals cover the area where false localizer paths may exist with strong 90- or 150-Hz modulation and cover up any false localizer paths. The modulation frequencies provide the correct information for returning to the approach path. These signals are called *clearance signals.*

Position Markers

ILS installations are provided with marker beacons that provide known fixes along the approach path. The system has the capability of providing three different marker beacons, the outer, middle, and inner, but the inner is seldom installed. The outer marker is placed between 3.5 and 6.5 nautical miles (nmi) from the touchdown end of the runway, the middle marker is 1000 feet (ft) from the end, and the inner marker, if it is used, is 500 ft from the touchdown end of the runway. All the marker beacons are placed on the extended runway center line. The marker system consists of ground beacon transmitters with an antenna pattern that directs the transmitter's energy directly overhead. A relatively insensitive receiver is mounted in the aircraft that can receive the marker's 75-MHz transmitter signal only when in the beam of the antenna.

The marker beacon transmitter is amplitude modulated with a pulsating tone at either 400, 1300, or 3000 Hz. These tones were chosen for several reasons. First, they are all in the speech frequency range. Thus the tones can be easily heard in the headphones and cockpit speakers that are optimized for speech bandwidth. Second, the harmonics of any one tone do not correspond to any other tone. This prevents harmonic distortion from one tone from causing a false triggering of another tone.

Photo 6–2 The MHz outer marker and compass locator. The 75 MHz antenna is on the roof of the house using the roof for a counterpoise while the low-frequency compass locator antenna is suspended from the utility pole.

The marker beacon receiver provides two functions. First, the receiver is connected to a speaker in the cockpit so that the pulsating tones can be heard by the flight crew. Second, an indicator lamp is illuminated on the instrument panel, designating the marker as outer, middle, or inner.

The marker beacon receiver can be one of two designs. The usual superheterodyne topology may be employed, but an interesting design uses the old tuned radio frequency (TRF) technique. The TRF topology is simply an RF amplifier and some amount of tuned circuits to provide the desired selectivity, as shown in Fig. 6-3. The TRF was used as an example of a bad radio receiver topology to describe the advantages of the superheterodyne in Chapter 4. The marker beacon receiver is required to receive only one frequency from a transmitter only a few thousand feet distant. This does not require a high-performance receiver, and the simplicity of the TRF may be exploited.

The marker beacon receiver uses a band-pass filter between the antenna and the RF amplifier, which is the only selectivity in the entire receiver. The RF amplifier provides the receiver gain and the gain-controlled element for the AGC system. A conventional AM detector is used after the RF amplifier. The dc component of the detected output is used to control the gain of the receiver while the ac component provides the audio output.

The detected output of the receiver feeds three band-pass filters at the three marker beacon modulation frequencies. Each filter receives the demodulated audio from the receiver and each output is rectified. The rectified output drives an indicator lamp. The marker beacon receiver has no flag indicator.

Many marker beacon receivers have a high/low sensitivity switch to reduce the sensitivity of the receiver. The greatest precision of a position marker is achieved when the marker receiver has the least sensitivity. This is because the less sensitive receiver will receive the marker only when directly overhead. On the other hand, the danger in using a low-sensitivity receiver is that the marker transmitter may have low power or the aircraft is

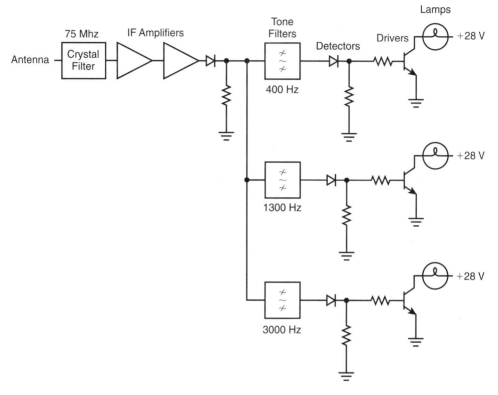

Figure 6–3   TRF form of marker beacon receiver.

not directly on the extended runway center line and the position marker is not received at all. The sensitivity switch allows the marker to be received, the switch changed to the low-sensitivity position, and the marker received again when the aircraft is directly overhead.

The greatest challenge when designing a marker beacon receiver is to prevent harmful interference from strong television signals. U.S. channel 4 extends from 66 to 72 MHz, while channel 5 extends from 76 to 82 MHz. It is permitted for both channels 4 and 5 to be assigned to a city, and thus it is possible for strong television signals to be present just below and above the 75-MHz marker beacon frequency. Because the input filter is the only source of receiver selectivity in the TRF design, it is necessary that the filter have sufficient rejection of strong television transmissions. Because of this requirement, some TRF marker beacon receivers use a high-performance crystal filter as the input filter.

Operation of navigation equipment at frequencies near high-power broadcast services is not unique to the marker beacon. The localizer, just discussed, has 108.1 MHz as its lowest operating frequency. Normally, this frequency is used as a test frequency, but some operating ILS facilities use this frequency. The highest FM broadcast frequency is 107.9 MHz, which is only 200 kHz removed. Broadcast transmitters are permitted to operate at very high power levels, and broadcast transmitters are often found on the outskirts of the city, along with airports. This situation has caused some serious problems with ILS reception. When

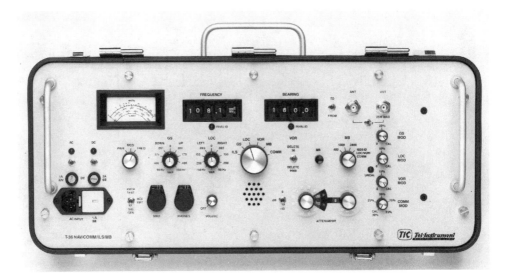

Photo 6–3 The Tel-Instrument T-36 test set may be used for ramp or bench testing of VOR, ILS, marker beacon, and communications equipment. Photo courtesy of Tel-Instrument Electronic Corp.

a city has a broadcast station operating on 107.9 MHz, prudent planning would avoid a low-frequency ILS assignment. Another solution is improved ILS receivers or the installation of a microwave landing system at the airport.

The Glide-slope System

Glide-slope guidance is provided using a technique very similar to that used for the localizer. An antenna pattern is provided by a carrier modulated with 90 Hz and the same carrier modulated with 150 Hz so arranged that the patterns merge and provide identical amounts of modulation on the desired glide path. Above the desired glide path, the 90-Hz modulation prevails, while the 150-Hz modulation prevails below the glide path.

The glide-slope operating frequencies are in the UHF spectrum from 329.15 to 334.70 MHz, with 0.15-MHz channels. For each localizer frequency, there is a paired glide-slope frequency, and thus there are 40 glide-slope frequencies. The frequencies paired with 108.10 and 108.15 MHz are, like their paired localizer frequencies, reserved for testing purposes.

A dedicated glide-slope receiver is used for receiving glide-path information. This receiver is not capable of being manually channeled to the glide-slope operating frequency, but is automatically channeled by the navigation receiver. This is an important characteristic since disastrous conditions would be generated if the glide slope and the approach path were for different approaches.

The typical glide-slope receiver is a superheterodyne using a synthesizer for the local oscillator. Because the range of input frequencies is relatively narrow, 5.2 MHz, and the

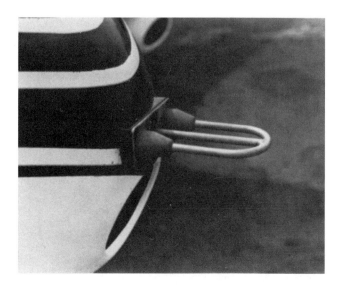

Photo 6–4 A glide-slope receiver antenna mounted on the nose of a light twin aircraft.

channel separation is relatively wide, 150 kHz, a fixed tuned RF input filter may be used. The IF can be in the region of 20 MHz or so, and the required selectivity can be obtained with an LC IF filter. A more-expensive crystal IF filter would be a better choice, and most glide-slope receivers use an IF of about 21.4 MHz with a crystal IF filter.

The demodulated output from the glide-slope receiver is fed to a signal converter essentially identical to the localizer converter. The major exception is that the glide-slope converter drives a horizontal meter pointer. The glide-slope indicator provides the same type of correction information as the localizer. Also, like the localizer, the glide slope generates a flag that is independent of the localizer flag.

RADIO ALTIMETER

As good as the glide slope system is, a more accurate system for assuring altitude control, the *radio altimeter,* is used for instrument landings. This device transmits a radio signal that is reflected from the ground and measures the time delay to the reflected return signal. Some early radio altimeters attempted to use a pulse type of transmission, which is difficult because the typical time delays are very short, only on the order of a few nanoseconds when the aircraft is near touchdown.

The technique that is employed is a frequency-modulated carrier for which the frequency is changed at a regular rate, as shown in Fig. 6-4. In this example, the frequency changes at a constant rate, and the difference between the transmitter frequency and the reflected frequency is a function of the distance. Consider the part of the frequency-modulation ramp where the frequency is increasing, and let t equal the elapsed time for the round trip from the altimeter to the ground and return. The reflected signal represents what was

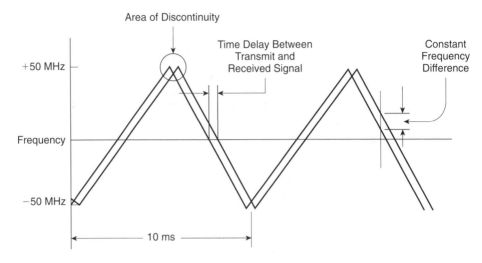

Figure 6–4   Frequency-modulation characteristics of a radio altimeter.

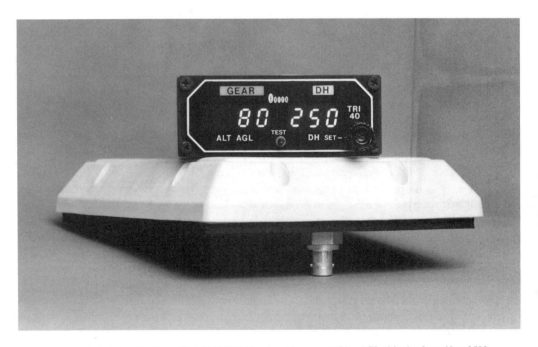

Photo 6–5   The Terra TRA 3000/TRI 40 radar altimeter provides AGL altitudes from 40 to 2500 feet and provides a "gear up" warning and decision height indication. Photo courtesy of Terra Corporation.

transmitted t seconds ago. If the transmitter frequency is constantly rising, the reflected signal is a lower frequency than the transmitter frequency. Let us assume that the frequency rise is r hertz per second (Hz/s). Therefore, the frequency difference is rt and is proportional to the altitude.

The frequency cannot rise at a constant rate forever and must be reversed to prevent the frequency from going to infinity. In the radio altimeter, the frequency rises and falls with exactly the same rate, but with a positive and negative sign, respectively. The radio altimeter takes the difference frequency between the transmitter and the reflected signal without respect to the sign and thus, rising or falling, the difference frequency is the same.

There is an area where the difference frequency is not a constant during the time that the frequency modulation changes slope, which is easily seen from Fig. 6-4. The period of nonconstant frequency difference is the same as the time delay but occurs twice each modulation cycle. If the period of the modulating waveform is long relative to the time delay, this period of nonconstant frequency difference will be very short relative to the entire period and may be disregarded. A typical modulation period is 0.01 s, which is a modulation rate of 100 Hz. Typical time delays are a few microseconds maximum. The frequency used for the radio altimeter is 4.3 gigahertz (GHz). The amount of frequency modulation is $\pm$ 50 MHz. The amount of error is greater at higher altitudes, but this is where a greater error can be tolerated.

Example

A radio altimeter uses a 100-Hz sawtooth modulation waveform as shown in Fig. 6-4 and a peak deviation of 50 MHz. What would be the difference frequency at an altitude of 200 ft?

Solution. The first task is to convert the 200 ft to meters, which is 61 meters (m). Second, the time delay is the time it takes the signal to travel the 122-meter round trip or, $t = 122/3 \times 10^8 = 407$ nanoseconds (ns).

The total frequency deviation is 100 MHz, because the frequency deviates $\pm$ 50 MHz from the nominal for a total of 100 MHz. The entire frequency change occurs in one-half the modulation cycle or 0.005 s. Thus, the rate of change of the frequency modulation, or r, is 100 MHz/0.005 s = 2×10^{10} Hz/s.

The frequency difference is $rt = 2 \times 10^{10} * 407 \times 10^{-9} = 8140$ Hz.

This example shows a period of 814 ns where the frequency difference is not a constant, which is insignificant compared to the total period of the modulation, which is 10×10^6 ns, or roughly one part in twelve thousand.

A block diagram of a radio altimeter is shown in Fig. 6-5. The transmitter is a frequency-modulated oscillator that also serves as the receiver's local oscillator. The transmitter power is routed to the antenna through a circulator. A *circulator* is a device in which power travels in only one direction.

The power from the oscillator is fed to the antenna, from which all the transmitter power is radiated. Receive power comes from the same antenna and travels around the circulator to the receiver mixer. The mixer multiplies the return signal with a signal from the transmitter oscillator, and the sum and difference frequencies are generated as in any superheterodyne receiver. The sum frequency is about twice the carrier frequency at 8.6 GHz. The difference frequency is a low frequency and is separated by the use of a low-pass filter.

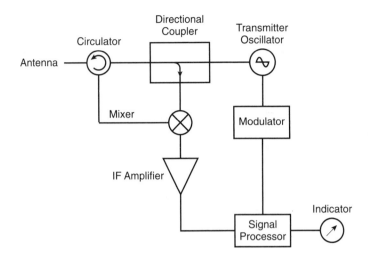

Figure 6–5   Block diagram of a radio altimeter.

The low-pass filter after the receiver mixer separates the difference frequency and also sets the maximum altitude at which the altimeter may operate. Usually, altimeters do not operate above about 2000 ft, which is a difference frequency of 81.3 kHz.

After the low-pass filter, an IF amplifier is provided to increase the signal strength to a level that can be used for digital circuits. The gain of this IF amplifier tends to favor the higher difference frequencies, which represent the return signals from the higher altitudes where the received signal is the weakest. Since this is an FM system for which only the frequency is important, a limiting amplifier is used and no AGC is required.

Since the altitude is directly proportional to the difference frequency, the difference frequency can be counted and used for a digital display. Many radio altimeters are analog instruments and thus the difference frequency must be converted to a voltage. This is done in a simple frequency-to-voltage converter.

ILS Accuracies

There are a number of methods of making an instrument approach to an airport. These include nondirectional beacon, VOR, and GPS. At the time of this writing, these are all *nonprecision* approaches. It is possible that GPS approaches will be certified for precision approaches in the future. A *precision* approach can be flown by the use of an ILS or an MLS. It must be understood that a precision approach using the ILS requires all components of the ILS to be employed. Therefore, the aircraft must be equipped with functioning localizer, glide-slope, and marker beacon receivers to perform a precision approach. The airport navaids must also be properly functioning. If one component is not operating, for example one of the marker beacons, the remaining components may be used for a nonprecision approach.

An ILS is capable of providing a certain level of accuracy, which depends on a number of factors. First, the capabilities of the ILS ground equipment are an important part of the overall system accuracy. The performance characteristics of the ILS transmitters, that is, modulation accuracy, lack of distortion, precise frequency control, and the like, are important. The accuracy of the transmitting directional arrays and the lack of sources of multipath are two items that will ensure that an accurate localizer and glide-slope signal environment will be provided. The frequency and accuracy of the maintenance checks of the ground equipment will have an effect on the overall accuracy. The longer the time between maintenance checks is, the more that the system tolerances can degrade before a check is made.

An accurate instrument approach and landing depend not only on the performance of the ground segment of the ILS but also on the airborne equipment. A properly installed system using high-quality equipment capable of high accuracy is required for accurate approaches and landings. It, too, must have a maintenance program.

Another component of an accurate approach and landing that may not be obvious is the skill level of the flight crew. Competence is assured by performing instrument landings on a regular basis, either in a real airplane or with a simulator.

There are three basic categories of instrument landings based on the level of performance of the three segments of an instrument landing: ground equipment, airborne equipment, and aircrew.

The category of a landing is determined by the ceiling and the visibility, which are defined as the *runway visual range,* or RVR. The least restrictive category is CAT I which allows instrument landings with an RVR of 2600 ft and ceilings of 200 ft. CAT II landings may be made with an RVR of 1200 ft and a ceiling of 100 ft.

There are three subcategories of CAT III landings. The first, CAT IIIa, allows an instrument landing with an RVR of 700 ft and a 50-ft ceiling. CAT IIIb allows an RVR of 150 ft and a ceiling of 35 ft. The ultimate, CAT IIIc, is for an RVR and ceiling of zero. Of course, in some larger aircraft, CAT IIIb, with its 35-ft ceiling, is as good as zero-zero.

Microwave Landing System

The venerable ILS, which has been in operation for over 50 years, provides one approach path, straight in, and one glide path, usually at 2.5°. While this may be suited for a large airliner, many situations would benefit from a steeper glide path or a curved approach. Other problems with the ILS are related to multipath propagation, which prevents many installations from being certified to CAT II or CAT III levels.

As you may recall, VOR has a multipath problem associated with the use of a directional antenna. One solution for this problem is the use of a different type of ground station, the Doppler VOR, which does not use a directional antenna. In the case of the localizer system, there is no alternative to the directional antenna. At some airports, typically those near mountains, it is not possible to install a reliable ILS. In some cases an ILS may be installed, but it will never be capable of CAT II or CAT III performance.

An improved landing system, called the microwave landing system, was developed and is being installed at various selected locations to allow more complex approaches and CAT II and CAT III approaches where the ILS could never provide this level of service. The flexibility of more complex approaches will improve traffic flow at congested airports.

The *microwave landing system* (MLS) is a *time reference scanning beam* (TRSB) system and uses an electronically scanned antenna beam transmitting carrier frequencies around 5 GHz. There are two sweeps of the antenna with a pause between the sweeps. The two sweeps are called the TO and FRO scan. The MLS receiver measures the time between the two sweeps, subtracts the pause time, and calculates the radial position of the aircraft.

The sweeping antenna patterns provide *angular guidance,* of which several are available from an MLS installation. First, the approach path or horizontal angular guidance, called *azimuth,* is required, which is the function provided by the localizer in the ILS. The vertical angular guidance, called *elevation,* provides the glide path and performs the function of the glide slope in the ILS.

The MLS may provide other angular guidances, such as flare, back azimuth, or a high-rate approach guidance. Which guidances are provided is a function of the airport. The complexity of the MLS installation depends on the requirements of the airport.

An important characteristic of the MLS should be noted. Since measuring the time difference between the TO and FRO scans of the MLS guidance transmitter allows any angle to be measured, any glide path or approach path can be chosen. This is in contrast to the ILS, for which the approach and glide paths are fixed.

MLS guidance signals are transmitted on one frequency that is unique to the MLS installation. At large airports, where there may be two MLS installations, each installation will have a unique operating frequency. There are 200 MLS frequencies available for use, which allows a much greater number of installations than ILS, which only has 40 frequencies. The guidance functions are time shared to allow the use of the same frequency. Essentially, each MLS transmitter takes turns transmitting its guidance information. There are some priorities, however. The more critical guidance signals, such as flare or approach elevation, must be transmitted more often than back-azimuth guidance or approach azimuth. Back azimuth is similar to the back course in an ILS installation.

As an example, for an MLS installation that has approach azimuth, approach elevation, and back-azimuth guidance, the most critical, approach elevation, is transmitted 39 times each second, while the approach azimuth is transmitted at one-third that rate or 13 times per second, and the back azimuth is transmitted only 6.5 times each second. Therefore, the approach elevation is transmitted three times, then there is one approach azimuth transmission, followed by three more approach elevation transmissions, and so on, as shown in Fig. 6-6.

Each MLS transmission is identified as to its function. This identification is achieved by transmitting a preamble, which is digital data. Several data are contained in the preamble, but the most important information is the guidance function that will follow. Immediately after the preamble, the TO scan begins, followed by the FRO scan.

Most MLS installations will also have a DME. The DME associated with the MLS is capable of more precision than the DME used with ILS installations. This improved DME is called a *precision DME,* or PDME.

The PDME operates using techniques similar to those of the conventional DME. The ground portion of the PDME is a transponder, and the airborne interrogator transmits two interrogation pulses and receives a reply. The major difference is that the rise times of the DME interrogation and reply pulses are very carefully controlled so that a precision time

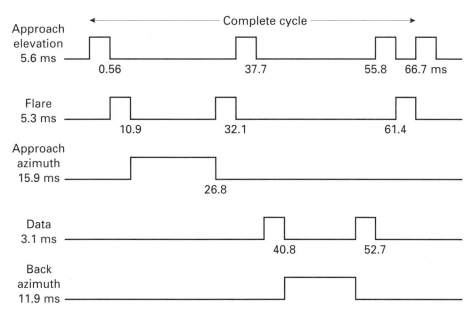

Figure 6–6   Sequence of transmission of an MLS installation having approach elevation, approach azimuth, and back azimuth.

reference may be obtained using only a small portion of the pulse rise time. This is in contrast to the conventional DME, for which the point of reference for timing was the 50% point of the DME Gaussian pulse, which has a pulse width of 3.5 microseconds (μs) and a rise time of 2.4 μs. Using these partial rise time measurements, the PDME is capable of measuring distance to within about 50 ft.

   The precision DME is used only for the very last portion of an MLS approach. The PDME located at an MLS installation serves all aircraft within the service area of the MLS installation, which could be a significant number of aircraft, particularly at a large airport. Because the PDME is critical for those aircraft making the final approach, the PDME has two modes of operation. The first mode is the initial approach or IA mode. This occurs between 7 and 22 nmi from the touchdown point. In this area, the PDME service is similar to the conventional DME, and the timing is referenced to the half-amplitude point of the Gaussian waveform. There is no need for 50-ft accuracy at 7 nmi or greater. When the aircraft has reached 7 nmi, the interrogator switches to the final approach or FA mode, and the PDME timing is relative to the partial rise time and the potential accuracy is increased to 50 ft. In addition, aircraft that are interrogating the PDME in the FA mode have precedence over those on IA mode.

TESTING LANDING SYSTEMS

The need to test radio navigation systems is evident. Flying IFR with faulty radio navigation equipment can be fatal, which has resulted in the testing of some radio navigation equipment on a regular basis mandated by law.

Two types of testing procedures may be employed. First, the navigation equipment may be removed from the aircraft and tested in a repair shop, a type of testing called *bench tests,* which refers to the workbench where the equipment will be placed while testing.

In the second method of testing, the equipment remains in the aircraft. This is called *ramp testing,* with the ramp referring to where the aircraft will be placed while the tests are performed.

Each method has its advantages and disadvantages, and debate rages between avionics professionals as to which method is superior.

The removal of the equipment from the aircraft to perform bench tests leaves much of the navigation system in the aircraft. As an example, if a VOR receiver is removed from the aircraft to test, the indicator, the antenna, the source of power, and all the interconnecting wiring remain in the aircraft. Electronic navigation equipment in modern aircraft consists of complete systems of components. If a navigation problem should exist, it could be a problem isolated to a single component of the system or a system problem. Problems could also be caused by other equipment that has nothing to do with the failed system.

If a system is properly specified, the performance of each electrical interface is fully characterized, and if all components of a system are tested fully, it is certain that the system will perform properly. This is the philosophy behind ARINC specifications. However, designing to ARINC specifications is expensive and few general aviation aircraft contain ARINC units.

An advantage of bench testing, however, is that the equipment may be subjected to more thorough testing. This is because inputs and outputs that cannot normally be accessed in the aircraft are available on the repair bench. Special test signals may be injected from sophisticated computer-generated tests to isolate failures. This is an important advantage because complex navigation equipment has literally hundreds of operational modes that cannot be investigated within an aircraft on the ground.

Another advantage is that tests can be performed over a longer period of time because an airplane is not disabled for the duration of the test. Ironically, some of the more sophisticated test systems begin to approach the price of a small airplane, and tying up the test system may have its own disadvantages.

What is the answer to the ramp–bench test debate? Both types of tests should be used. When a major evaluation of a navigation system is needed, a bench test should be performed on each element of the navigation system, and after the equipment is installed in the aircraft, a system ramp evaluation should be performed.

Test Equipment

A number of test equipment types are used to evaluate avionics systems. The most common and important are signal simulators, which can provide simulated VOR, localizer, glide-slope, marker, DME, and other types of radio navigation signals. Standard signals may be provided as well as degraded signals to check for the operation of flag circuits.

Usually, a signal simulator can provide any standard operating frequency, with signal levels from less than a microvolt to hundreds of millivolts to simulate very close and very distant navigation aids.

There are a number of performance standards against which a piece of equipment may be tested. In some cases, the level of performance is specified in the Federal Aviation Regulations. Usually, the actual performance specifications are not in the FARs, but the FARs refer to a TSO. One notable exception is the air traffic control transponder, both ATCRBS and mode-S. The actual performance levels for an installed system are listed in part 43, appendix F of the FARs, in spite of the fact that the FARs require that the transponder be TSOed.

If the equipment to be tested is not TSOed, then the manufacturer's specifications are the performance limits that judge whether a piece of equipment has passed the tests or not.

There is a ramp test method that permits pilots to test various types of navigation systems without the use of ramp test equipment; it is called the VOR test facility, or the VOT. This facility is a permanently installed signal simulator at an airport that radiates a low-power signal that can be received on the airport property. Unlike a real VOR, the VOT simulates the signal without the use of a rotating antenna pattern, and thus the zero-degree TO test signal is not affected by the position of the aircraft. The facility's name implies that it can only test VOR equipment, but this is not the case. Some VOTs can test VORs and DMEs. Using a VOT as a test of the VOR receiver in an aircraft satisfies the equipment testing requirements of the FAR for IFR flight.

REVIEW QUESTIONS

6.1. What is the localizer DDM at the runway center line?

6.2. What is back course? How does the use of back course signals differ than the normal approach course?

6.3. What are clearance signals? Why are they used?

6.4. What position markers are used? On what frequency do they transmit?

6.5. What carrier frequency does the position marker use? What other radio services use frequencies near the marker?

6.6. What modulation frequencies are used for the position markers? What characteristic do the modulation frequencies have that prevents interference?

6.7. What frequencies are assigned to the localizer?

6.8. What differentiates a localizer ident from a VOR ident?

6.9. What frequencies are used for the glide slope?

6.10. How is the glide-slope receiver frequency selected?

6.11. What type of Morse code modulation is used for identifying the glide-slope facility?

6.12. What landing systems use DMEs?

6.13. What distinguishes PDME from normal DME?

6.14. How does an MLS receiver know which MLS service is being received?

6.15. Why are some MLS transmissions made more often than others?

7

Aircraft Surveillance Systems

CHAPTER OBJECTIVES

In this chapter the student will learn about the major radar-based surveillance systems and the function of the ground and airborne components. A system understanding of the TCAS collision avoidance system will be gained.

INTRODUCTION

Radar is the subject of this chapter. Aircraft surveillance is the application of various types of radar systems. *Radar* is an acronym for RAdio Detection And Ranging. Radio signals are transmitted in a very narrow beam, and the reflected signals are received from *targets*. The direction of the antenna determines the direction to the target, while the elapsed time between the transmission of the signal to the receipt of the reflected energy is used to determine range. Radar was developed during World War II as an important military tool and much of that work is in use today for both military and civilian radar applications.

Although the concept of radar sounds simple, there are significant practical problems to be solved. The reflected energy is from irregular bodies such as aircraft fuselages. The reflections are in a number of directions, and only the energy that is reflected in the direction of the radar antenna is received. The phenomenon of reflection in all directions is called *scattering*. The energy that is scattered in the direction of the radar receiver is called the *backscatter*. The ratio of the energy falling on a target to that which is backscattered is called the *radar cross section*. If the wavelength of the radar transmitted energy is long compared to the dimensions of a target, energy is not scattered but simply goes around the target.

Even for relatively small wavelengths, a small amount of energy is backscattered from the target.

When a radio signal is radiated, the energy density, that is, the energy per unit area, continually decreases. This is because, as a radio signal travels away from the transmitter, the energy is spread out over an increasingly larger area. The energy density decreases as the square of the distance from the transmitter. When energy is backscattered, the process is as if the target radiated the signal, and the energy spreads out over an increasingly larger area in the same fashion as when the signal was transmitted. Therefore, when the backscattered signal is received at the radar, the signal level has decreased as the distance to the fourth power. This presented the first practical problem to the developers of radar. The levels of backscattered return signals were very weak, which required a transmitter with high-power, short-wavelength energy.

The position of the target is a function of the time of flight of the radar return and the direction of the antenna. To achieve reasonable resolution, a narrow-beamwidth antenna is required. This was the second of many practical problems.

There were a number of other difficult problems to be worked out, and a very large group of researchers was assembled at the Massachusetts Institute of Technology, as well as other research laboratories around the world, to provide practical solutions to these problems.

AIR TRAFFIC CONTROL TRANSPONDER

Aircraft surveillance for air traffic control is accomplished using two types of radar. A primary radar determines an aircraft's location by transmitting a very high power pulse of RF energy and receiving energy reflected from the aircraft. This type of radar does not require the installation of special equipment in the aircraft being detected. However, only very basic information concerning the detected target is known. The strength of the radar return signal gives some indication of the size of the aircraft, but this is not totally reliable. No information concerning the identity of the aircraft is available.

The strength of the return signal is a function of the radar cross section of the aircraft. The cross section is a measure of how much energy is reflected by the target to the radar, which is a function of how large the aircraft is and how well the aircraft reflects. Some aircraft are deliberately designed to have a low radar cross section in spite of the fact that the aircraft is relatively large; there are military "stealth" aircraft. The military must also know the identity of the aircraft seen by radar, because both friendly and hostile forces produce the same radar blip.

During World War II, the need for positive identification of a radar target was known and a system called a *secondary radar* was invented. The original purpose of the secondary radar was to provide an "identification of friend or foe," or IFF, function. Providing identification of an aircraft is equally important to air traffic control radar, and after the war, the IFF system was expanded to include civilian applications. This expanded system was called the *air traffic control radar beacon system,* or ATCRBS.

Secondary radar uses radar transponders in the aircraft that are to be identified. The transponder contains a transmitter and a receiver. When a transponder receives an interrogation from a ground transmitter, it replies with certain requested information by transmitting that information back to the ground station.

The ATCRBS transponder served the aviation industry well for many years, but increased air traffic began to strain the system in the 1980s. The problems were primarily frequency congestion from too many transponders and a situation called garble. Because of this, a new system was developed and installations began in the late 1980s. The new system has the capability of a ground station addressing aircraft individually, so a transponder will reply only when it is specifically addressed. This system is called the mode-S system, where the S refers to the ability of the system to selectively address transponders.

Garble occurs when two or more transponders are at nearly the same distance from a ground radar site and close enough so that more than one is in the beam of the antenna. Therefore, when the interrogation is received by the transponders, the replies will arrive at nearly the same time at the interrogator and mutually interfere.

Although mode-S transponders and interrogators are being installed, the older ATCRBS transponders and interrogators will be operational for a long time.

The ATCRBS system uses an interrogator colocated with the primary radar. The secondary radar uses a directional antenna that rotates with the primary radar and interrogates the transponder in the aircraft at a frequency of 1030 megahertz (MHz). The transponder, upon receiving the interrogation, replies at a frequency of 1090 MHz with the information requested. Pulse modulation is used with the ATCRBS transponder system. Short pulses of RF energy are transmitted to the aircraft, from which a pulse-amplitude-modulated reply is returned.

Two types of information may be requested from an ATCRBS transponder, identity and altitude. The type of information desired is encoded in the interrogation. An ATCRBS interrogation consists of two pulses; the time between the two pulses designates the type of interrogation. A third pulse is included in an interrogation to prevent replies from the side lobes of the antenna, which will be described later.

If the identity of the transponder is desired, two interrogation pulses, P_1 and P_3, separated by 8 microseconds (μs) are transmitted in what is called a mode-A interrogation, as shown in Fig. 7-1. The transponder recognizes this interrogation and transmits its identity. If the interrogator requires the altitude, the interrogation consists of P_1 and P_3 separated by 21 μs, which is a mode-C interrogation and produces an altitude reply. Do not confuse the mode-A reply with the altitude reply, which is mode-C. It is unfortunate that the A of the mode-A reply is often confused with altitude.

The secondary radar interrogator uses a directional antenna; as for all directional antennas, the antenna pattern includes a main lobe, which is the forward direction of the antenna, as well as side and back lobes, as shown in Fig. 7-2.

Although most of the energy radiated by the antenna is from the main lobe, a significant amount of energy is radiated from the side and rear lobes. An aircraft that is close enough to the interrogator may receive sufficient energy from a side lobe that it would reply. When a ground interrogator is capable of interrogating a transponder from a side lobe, it is also able to receive the reply, which in this case is from the side of the antenna. Expecting the reply from the main lobe, the ground interrogator would interpret the aircraft position incorrectly.

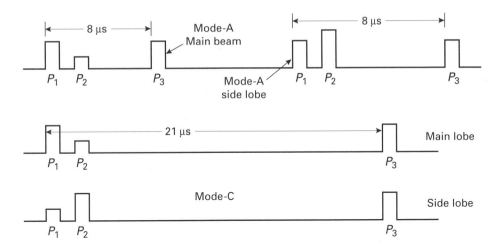

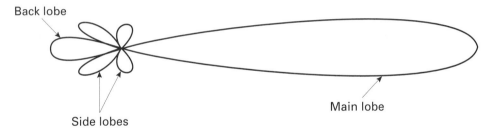

Figure 7–1   Mode-A and mode-C interrogations from the main and side lobes of the interrogator antenna.

Figure 7–2   Radiation pattern of a typical directional antenna.

A third pulse, P_2, is transmitted from an omnidirectional antenna such that the signal strength of the omnidirectional transmission is greater than that of any side or back lobe and less than the main lobe. P2 is transmitted 2 µs after the first interrogation pulse, as shown in Fig. 7-1. This pulse is called the *side-lobe suppression,* or SLS, pulse. The transponder uses this omnidirectional pulse as a comparison to the level of the interrogation pulses. If the level of the SLS pulse is less than the interrogation pulses, the interrogation is received from the main lobe and the transponder is free to reply. If the SLS pulse has a greater amplitude than the interrogation, the interrogation pulses are from a side or back lobe, and the transponder does not reply.

As in the primary radar system, the secondary radar system determines the range to a target by measuring the elapsed time from the transmission of an interrogation to the receipt of a reply. Of course, there must be a finite reaction time of the transponder to the interrogation, which is a part of the total delay time. To ensure an accurate elapsed time measurement, the delay time must be standardized and well controlled. The delay time is 3 µs as measured from the P_3 interrogation pulse to the first framing pulse, F_1, of the reply.

Example

What is the total delay time as measured from P3 of the interrogation pulse to the first framing pulse for an aircraft 10 nautical miles (nmi) from the interrogator?

Solution.   The total delay time is the transit time, which is 10 nmi × 12.35 μs/nmi, plus the internal delay time, which is 3 μs, or 126.5 μs.

When a transponder recognizes a valid interrogation or side-lobe interrogation, it will not receive any further interrogations for 50 μs to prevent the transponder from replying to a multipath interrogation. Seldom do multipath interrogations arrive more than 50 μs delayed from the desired interrogation. If the transponder has received a valid interrogation, it will transmit a reply, which implies that for more than half of the 50-μs suppression period the transponder receiver is suppressed due to the fact that the transponder is transmitting.

The modulation of a transponder reply is pulse amplitude modulation or, more accurately, pulse existence modulation. If a logic 1 is to be transmitted, a pulse exists in a time slot, and if a logic 0 is to be transmitted, no pulse exists.

Every reply starts with a pulse, which is the first framing pulse F_1, as shown in Fig. 7-3. This pulse is the reference for the time slots of the reply. Every 1.45 μs after the first framing pulse, there is a time slot. A second framing pulse follows the last data pulse. The two framing pulses help the ground interrogator to recognize a valid reply. Remember that pulse existence modulation is used, and should the reply be all zeros, only the two framing pulses will be transmitted. Having two framing pulses makes it easier to recognize a reply. There are 15 time slots of which 14 are used. The seventh time slot is blank and never contains a pulse, and the first and last time slots are for framing pulses, leaving 12 time slots for data.

Twelve data pulses of binary information can be used to transmit up to 4096 different identities, which is not enough to allow every aircraft in the world to have a unique identity code. Therefore, the codes are recycled, and an aircraft that is in controlled flight must obtain an identity code, called a *squawk,* from the air traffic controller. There are some generic codes such as 1200, which indicates an aircraft is in VFR flight, and 7700, which denotes an emergency situation. These codes are used when necessary and not obtained from air traffic control.

The identify code is entered into the transponder using octal digits. Octal digits span from 0 to 7, which can represent 3 binary bits. Thus, to represent 12 binary data bits, which is required by the transponder, four octal digits are used. Therefore, the transponder codes span from 0000 to 7777, which represent a binary 0000 0000 0000 to 1111 1111 1111. Figure 7-4 shows an example of the pulse sequence for the squawk 1234.

The altitude of the aircraft is transmitted using the same 12 data time slots using a unique code called the Gray code. The *Gray code* is a special binary code in which only

Figure 7–3   ATCRBS transponder reply sequence.

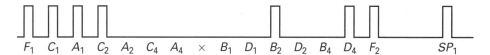

F_1 C_1 A_1 C_2 A_2 C_4 A_4 × B_1 D_1 B_2 D_2 B_4 D_4 F_2 SP_1

Figure 7–4   Example of a transponder code.

1 bit changes between subsequent codes. This characteristic is useful when mechanical encoding devices are used to reduce the amount of error due to mechanical problems. Problems tend to occur in a mechanically encoded input when a large number of bits are to change, such as the transition from 01111111 to 10000000. Since all 8 bits change, it is possible that not all the bits will change at exactly the same time and thus erroneous outputs will be produced. As an example, assume that the most significant bit changes late in the preceding example. Therefore, the encoder output will go from 01111111 to 00000000 and finally to 10000000. The situation is that the outputs go from 127 to 0 to 128, which produces an interim output that is in error by half-scale. In the case of the Gray code, only 1 bit changes; thus, if that bit changes too soon or too late, the error can only be one least significant bit. Mechanical misalignment can cause bits to change too soon or too late, and therefore the Gray code is used for mechanical encoders. The altitude of the aircraft is determined from a specially equipped altimeter called an *encoding altimeter.* The altimeter, being a mechanical device, benefits from the use of the Gray code. Figure 7-5 shows some representative Gray codes with only 1 bit changing between subsequent altitudes.

There is nothing to differentiate a mode-C reply from a mode-A reply. The interrogator only knows the difference because the reply is in response to either a mode-A or mode-C interrogation.

An additional pulse that occurs 4.35 µs after the second framing pulse is called a *special pulse identifier,* or SPI. When an air traffic controller assigns a squawk to an aircraft, to help identify that aircraft from the large number of aircraft displayed on the radar screen, the controller asks the aircrew to "identify." The aircrew then presses an ident button on the transponder, which adds the additional SPI pulse to the transponder's reply. This extra pulse causes the air traffic controllers's radar display to identify the aircraft transmitting the extra pulse by causing an increase in the brightness of the identified aircraft. In this fashion, the air traffic controller is certain of which displayed aircraft he or she is communicating with and that other aircraft have not mistakenly used the assigned squawk and identified. The added SPI pulse remains for about 15 seconds(s), after which an internal timer in the transponder removes the pulse.

Transponders can be interrogated by a number of air traffic control radar sites. To an interrogator, some signals received from transponders are not a result of interrogations. These replies, called *fruit,* are intended for other radar sites. This fruit must be removed by the ground interrogator so that the radar display is not cluttered with extra replies. A device called a *defruiter* is used to remove these signals that are not a result of an interrogation. The technique is simple. After an interrogation, the signals received at the ground station are stored in a computer memory along with the time of arrival as measured from the interrogation. A second interrogation is made, and the replies are again stored along with the time of arrival. This is done a number of times. Signals that were actual replies to interrogations

Examples of Pressure Altitude Reporting Codes

Altitude	D_2	D_4	A_1	A_2	A_4	B_1	B_2	B_4	C_1	C_2	C_4
0	0	0	0	0	0	0	1	1	0	1	0
100	0	0	0	0	0	0	1	1	1	1	0
200	0	0	0	0	0	0	1	1	1	0	0
300	0	0	0	0	0	0	1	0	1	0	0
400	0	0	0	0	0	0	1	0	1	1	0
500	0	0	0	0	0	0	1	0	0	1	0
600	0	0	0	0	0	0	1	0	0	1	1
700	0	0	0	0	0	0	1	0	0	0	1
800	0	0	0	0	0	1	1	0	0	0	1
900	0	0	0	0	0	1	1	0	0	1	1
1000	0	0	0	0	0	1	1	0	0	1	0
1100	0	0	0	0	0	1	1	0	1	1	0
1200	0	0	0	0	0	1	1	0	1	0	0
1300	0	0	0	0	0	1	1	1	1	0	0
1400	0	0	0	0	0	1	1	1	1	1	0
1500	0	0	0	0	0	1	1	1	0	1	0
1600	0	0	0	0	0	1	1	1	0	1	1
1700	0	0	0	0	0	1	1	1	0	0	1
1800	0	0	0	0	0	1	0	1	0	0	1
1900	0	0	0	0	0	1	0	1	0	1	1
2000	0	0	0	0	0	1	0	1	0	1	0
2100	0	0	0	0	0	1	0	1	1	0	0

Figure 7–5 Examples of the Gray code used for altitude reporting.

will appear practically every interrogation at the same elapsed time, whereas signals that just happened to arrive after an interrogation will not happen again at the same time, and new random signals will appear. Only those replies that occur repeatedly at the same time after an interrogation are displayed. To be absolutely certain that two interrogators are not synchronized, resulting in synchronized fruit, the interrogation rate of radar sites that are less than a few hundred nautical miles from each other are deliberately made different.

Extraneous replies due to multipath may be removed by the defruiter. In a multipath situation, the interrogation signal takes two or more paths. One path is obviously the direct and desired path. Additional paths involve reflections and are longer. The desired situation is that the interrogation arrives at the transponder via the direct path and that the reply travels the same path. However, when multipath propagation is present, the interrogation could arrive via two paths and the reply could travel the same two paths and results in a reply via the same desired direct path. Transponders are deliberately suppressed from replying to a interrogation that arrives sooner than 50 μs after the receipt of a valid interrogation to reduce the possibility of replying to a multipath signal. If the multipath signal arrives more than 50 μs after the valid interrogation, the transponder will reply.

If a transponder replies to two interrogations, one due to multipath, the two replies can, likewise, travel two paths to the interrogator. Thus, the interrogator will receive four replies. Of the four replies, two will mutually interfere at the ground interrogator. These two are the replies from the direct uplink and the multipath downlink interfering with the reply from the multipath uplink and the direct downlink. The desired exchange, direct uplink and downlink and the multipath uplink and downlink, will arrive without interference.

These false replies are easily recognized as they occur at the same fixed time delays after the desired reply and contain the same information.

The transponder is immunized from replying to a multipath interrogation by suppressing the transponder for 50 μs after the receipt of an interrogation. The 50-μs time delay is equivalent to an interrogation that traveled 8 nmi farther than the desired interrogation. In many cases, particularly in mountainous terrain, an 8-mile (mi) path differential is not uncommon. The transponder can transmit only one reply, but because of the multipath the interrogator can receive two replies.

The Transponder

The transponder contains a receiver and transmitter in which there can be some or practically no sharing of circuits. A typical transponder is shown in Fig. 7-6. The receiver is a common superheterodyne using a crystal-controlled local oscillator. A crystal oscillator operating at 107.77 MHz is multiplied by 3 to 323.333 MHz and again multiplied by 3 to 970 MHz. This provides a local oscillator for a 60-MHz IF.

The RF input is filtered to remove the image frequency, and an IF filter at a 60-MHz center frequency is provided at the mixer output. Transponder interrogation pulses have a bandwidth of about 2 MHz, and the appropriate IF bandwidth is on that order.

The IF amplifier provides the majority of the receiver gain. The IF amplifier used in secondary radar transponders is of a type called a *logarithmic amplifier*. This form of amplifier provides a detected output that is proportional to the logarithm of the input, which compresses a very large dynamic range to a smaller range of voltage.

The output of the log amplifier feeds the pulse amplitude detector. This circuit compares the amplitude of P_1 to P_2. If P_1 is equal to or greater than P_2 in amplitude, the transponder replies. If the amplitude of P_1 is 9 decibels (dB) greater or more than P_2, the transponder should reply. Between 0 and 9 dB, the transponder may or may not reply. Typically, the SLS detector has a nominal 4.5-dB threshold.

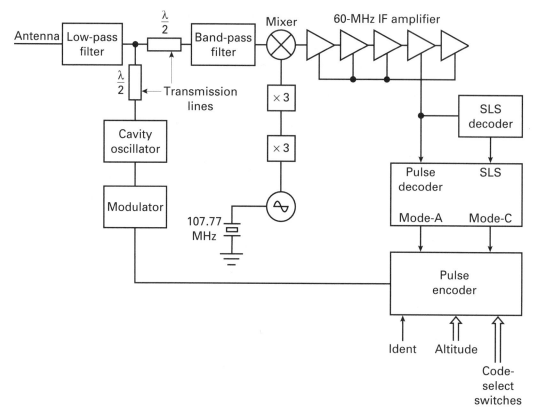

Figure 7–6　Block diagram of a secondary radar transponder.

　　The output of the IF amplifier also feeds the pulse decoder, which determines if P_1 and P_2 constitute a valid interrogation and whether the interrogation is a mode-A or mode-C. If a valid interrogation has been received, an output starts the time delay, which inserts additional time delay so that the total transponder delay is 3 μs.

　　After the time delay, the reply is transmitted. The type of reply, mode-A or mode-C, is determined by the type of interrogation received. The basic pulse positions and widths are the same for a mode-C or mode-A reply, but the source of data is different. The front panel switches provide the data for a mode-A interrogation, whereas the data from the encoding altimeter are used for a mode-C interrogation.

　　The mode-A/C decoder selects the source of transmitted data and the pulse modulated output is transmitted. This is usually a shift register that is parallel loaded with the desired data and clocked serially with a 1.45-μs clock.

　　For many years, the transponder transmitter was a single vacuum tube oscillator mounted in a resonant cavity. It is not possible to make a frequency-stable resonant circuit using conventional capacitors and inductors at the required frequency of 1090 MHz. A cavity resonator uses a cylinder that is a section of transmission line.

Although the vacuum tube transmitter served well, the performance of the ATCRBS system was limited by the performance of the vacuum tube cavity oscillators, primarily in the area of frequency stability. To improve the performance of the secondary radar system, the required frequency tolerance of the ATCRBS system was reduced for most classes of transponders to the point that the vacuum tube transmitter is no longer a viable alternative. However, because so many transponders used them, a significant number of vacuum tube transmitters will be present for many years.

New transponders use a transistorized transmitter in which a transistor oscillator is followed by several stages of amplification to isolate the oscillator from the antenna. A mismatched antenna could cause a significant frequency error with a vacuum tube transmitter. The inclusion of the amplifier stages prevents this effect from becoming a problem with the transistor transmitters.

The ATCRBS transponder operates on two frequencies in the L band, 1030 and 1090 MHz, and is not the only pulse-modulated L band system aboard an aircraft. The DME also operates with pulse modulation in the L band. Although the DME frequencies are different from the transponder frequencies, when two high-power L-band transmitters are located close to each other, it is difficult to prevent mutual interference.

A system called *mutual suppression* is employed that disables an L-band receiver so that the high-power transmitter does not cause the reception of a very distorted transmission, which could possibly disrupt the operation of a transponder or DME. The protected receiver is not operating and may miss an interrogation in the case of a transponder or a reply in the case of a DME. Both of these systems can operate successfully with occasional missing pulses.

THE MODE-S SYSTEM

The ATCRBS secondary radar system has been operational for more than 50 years and has served well. However, some weaknesses of the system have become critical as the density of air traffic increases.

Every transponder that receives a valid interrogation will reply. If two or more aircraft are within the beam of the interrogator antenna, these aircraft will reply to the same interrogations. If the aircraft are not exactly the same distance from the interrogator, the aircraft will reply at slightly different times, and the two replies will arrive as two distinct replies and clearly from two different aircraft. However, if the aircraft are relatively close, the two replies can overlap in time and produce what is called *garble*. No error detection is applied to transponder replies. When replies are corrupted, for whatever reason, the lack of error detection could cause an interrogator to accept a false response.

Just how close do aircraft have to be to cause garble? Without the SPI pulse, the length of a transponder reply is about 24 μs. However, if two replies occur so close together that it is not clear where one stops and the other begins, they cannot be decoded. For two replies to be separated so that there is no question that there are two replies, they must be separated by about 30 μs. To the interrogator, the difference in the arrival time of two replies is 12.35 μs/nmi, which is based on the concept of the radar mile. Therefore, the two aircraft must be separated by more than 2.4 nmi. This difference in distance is from the aircraft to

Photo 7–1   The Bendix/King TRA 67A ATC transponder showing the indicator LEDs for the built-in test function. Photo courtesy of Allied Signal Commercial Avionics System.

the interrogator. As an example, two aircraft separated by 2.5 nautical miles in altitude but at the same radial distance from the interrogator will still garble. In heavy air traffic, situations such as this are not uncommon.

The mode-S concept in secondary radar transponders was developed to eliminate the problem of garble in heavy air traffic. Rather than interrogations that allow any transponder to reply, transponders are individually addressed. Thus, if two aircraft are physically near, their transponders are selectively interrogated and only the addressed transponder will reply. Of course, a transponder cannot be selectively addressed until it is known what transponders are present, and there are provisions for causing all transponders to reply so that a roll call of transponders may be generated.

This improved transponder system is called *mode select* or mode-S. In this system, every aircraft has a unique identity because there are provisions for 16.8 million different aircraft to be uniquely identified. The mode-S system also contains a very powerful error-detection scheme so that only one interrogation will result in a trustworthy reply. Therefore, many fewer interrogations and replies are required for the operation of the system, which reduces congestion on the 1030- and 1090-MHz frequencies.

The mode-S system was designed to cooperate with the ATCRBS system so that two new operating frequencies would not have to be assigned, and once the mode-S transponders are installed, their appearance will enhance the existing system. The alternative would be to have two separate systems in operation until one is phased out. To do this, the mode-S transponder must behave like an ATCRBS transponder when interrogated by an ATCRBS interrogator, and ATCRBS transponders must not degrade the mode-S system.

There are two distinct types of mode-S interrogations, all-calls and selective. The all-call interrogations cause all transponders, ATCRBS and mode-S, to reply; this is shown in Fig. 7-7. However, an extra pulse, P_4, is added during the time when an ATCRBS transponder receiver would normally be suppressed, which is invisible to the ATCRBS transponder, but the mode-S transponder can decode it. This pulse has two possible widths, 0.8 and 1.6 μs. When the narrow P_4 is present, the interrogation is an ATCRBS-only all-call, and the mode-S transponder does not reply. When P_4 is a wide pulse, the interrogation is an ATCRBS/mode-S all-call and the mode-S transponder replies with a mode-S reply. Because the P_4 pulse is invisible to the ATCRBS transponder, it replies to both all-calls. Therefore, an all-call can be used to cause all mode-S transponders to identify. Once a mode-S transponder is identified, it can be selectively called and tracked.

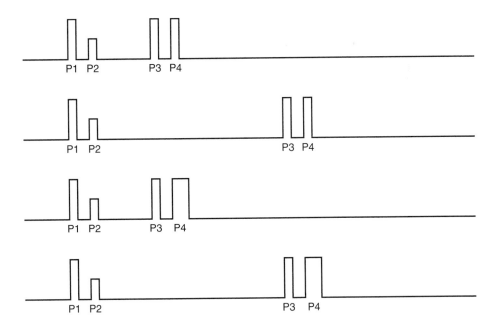

Figure 7–7   Mode-S all-call interrogations.

Considerably more data can be extracted from a mode-S transponder; however, the important data are, as always, identity and altitude. The mode-S transponder is also capable of spontaneous transmissions, unlike the ATCRBS transponder, which can only respond to interrogations. The mode-S transponder transmits squitter, which is unsolicited transmissions used in the TCAS system.

The mode-S transponder serves as the communications link for the collision avoidance system, TCAS. For this purpose, various data are exchanged, such as the airspeed capability and installed equipment of the mode-S equipped aircraft. The mode-S transponder can be used for numerous other data exchanges to aid air traffic control and other safety and traffic control functions. This is called the *data-link* capability of the mode-S transponder.

Selective interrogations require considerably more data than can be included in three or four interrogation pulses. For selective interrogations, a three-pulse interrogation is transmitted consisting of two pulses, P_1 and P_2, separated by 2 µs and followed by a long pulse, P_6, as shown in Fig. 7-8. This long pulse contains data encoded using *differential phase shift keying,* or DPSK. This form of modulation reverses the phase of the carrier for a 0.25-µs period. *Differential* implies that the phase of the carrier is reversed only if a logic 1 is to be transmitted. Each bit of the data word has a time slot during which the phase of the carrier may be reversed. Thus, each transmitted bit is referenced to the previous bit, which requires that a reference be provided for the first bit, which has no preceding bit. In addition, a time reference is required, which is used to locate the time slots used for data transmission. The first 1.25 µs of P_6 is a steady phase carrier that is used to synchronize the demodulators used in the receiver. A phase reversal is provided exactly 1.25 µs after the beginning of the pulse, which is called the *synchronizing* or *synch phase reversal,* to provide a time reference for locating the data time slots, which begin 0.5 µs after the synch phase reversal. Each data time slot is 0.25 µs in width, and P_6 contains either 56 or 112 time slots depending on the type of interrogation.

Of the 56 bits of information transmitted, only 24 contain the address of the interrogated transponder. The remaining bits are used for additional data to be transmitted to the transponder.

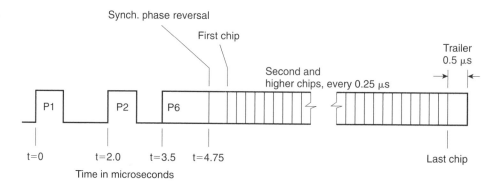

Figure 7–8   Mode-S selective interrogations.

Each interrogation contains an error-detection code called a *cyclic redundancy check,* or CRC. The CRC may be generated using some or all of the transmitted bits, but always involves the address of the transponder. In the ATCRBS transponder system, error detection is achieved by repeated interrogations and checking the replies for consistency. For this to be effective, considerably more replies must be transmitted by the transponder than absolutely necessary. The mode-S system was developed to alleviate congestion of the radio frequencies, and this can only be achieved by reducing the number of transmissions made by interrogators and transponders. It is desired that only one transmission be made for each transponder, which requires that the one and only interrogation and reply be known to be absolutely correct.

The CRC error check involves generating a set of bits called the CRC bits from the transmitted data. To generate a CRC, the data to be protected are treated as a binary number and calculations are made using that binary number. The data number is divided by another binary number, which is represented as a series of ones and zeros called a *generating polynomial.* The division, done with binary arithmetic, like any other division, provides a dividend and a remainder. The remainder is subtracted from the data, and if the difference is divided by the same generating polynomial, the result will be a dividend with a zero remainder. If the combination of data and CRC bits is corrupted, the division by the generating polynomial will not produce a zero remainder and the corruption can be detected.

In the mode-S system, a special form of simplified binary division is used, and the remainder is combined with the data by interleaving data and CRC bits. The data are checked by performing a binary division process, and the result is checked for a zero remainder. The ability of the CRC check to detect errors is very effective, and very few combinations of errors escape detection.

Each mode-S transponder has a unique address that is permanently programmed into the transponder. Unlike the ATCRBS system in which an air traffic controller assigns a squawk to be programmed into the transponder, the mode-S transponder is programmed by a unique wiring of the transponder's connector, the part that remains in the aircraft. Even when the transponder is removed and replaced with another unit, the new transponder responds with the unique code assigned to the airframe. Thus, each transponder is identified by the airframe in which it is riding, and an air traffic controller does not need to uniquely identify each aircraft. Every reply from a mode-S transponder is identified with the transponder's address and, like the interrogation, the mode-S replies contain a CRC error-detection code.

Also like the interrogations, the mode-S replies contain either 56 or 112 data bits corresponding to a short or long reply. Mode-S replies are transmitted on a frequency of 1090 MHz using pulse position modulation. The difference between the uplink and downlink modulation was chosen to make the design of the mode-S transponder transmitter easier. The choice of phase shift keying for interrogations saves spectrum, but requires the use of a stable transmitter frequency. For the ground interrogator, this stable transmitter does not constitute an undue expense. However, to require the airborne transponders to have such stable transmitters would add to the weight and cost of the transponder. Pulse position modulation was chosen for the mode-S replies from the transponder to permit a simpler transmitter design.

The mode-S transponder reply consists of a pulse position modulation transmission as shown in Fig. 7-9. Each pulse has a duration of 0.5 μs, and a 1-μs time slot is assigned to each transmitted bit. To transmit a logic 1, the pulse is transmitted in the first half of the 1-μs time slot, whereas a logic 0 is distinguished by transmitting the pulse in the latter half of the 1-μs time slot.

Each reply is preceded by a preamble that consists of two logic 1s followed by a missing pulse and two logic 0s. This is a unique pulse sequence; the missing pulse ensures that the preamble will never appear as a sequence of valid data pulses and will be recognized as the preamble. The preamble contains two leading edges on even 1-μs intervals and two leading edges on 1/2-μs intervals that the ground receiver can use to affect timing.

There are some important characteristics of this data transmission scheme. First, each reply has the same number of pulses and each pulse is the same width. The short reply has 60 pulses, the 56 data pulses and the 4 pulses from the preamble. The long reply has 116 pulses, regardless of the data transmitted. This ensures that the energy contained in a reply is the same regardless of the data transmitted. When the energy content is a function of the transmitted data, certain data content is more reliably transmitted than others. This was the case with the ATCRBS transponder. A reply of all zeros consists of only the two framing pulses, whereas a reply of 7777 plus an ident, SPI pulse, consists of 15 pulses that contains 7.5 times the energy of an all-zero reply.

Another useful characteristic is that there is a transition, that, is a pulse either being turned on or off, at the center of each time slot. This transition can be used to help establish the timing for the data slots. Again, this characteristic was not available in an ATCRBS reply. The only pulse that could be used for timing was the first framing pulse. The data pulses were not useful because they may not exist and the second framing pulse occurs after the timing is done.

Mode-S transponders are typically interrogated only once. If a reply is received from an interrogation, because the security of the data transmission is assured by the CRC bits, only one reply constitutes a successful interrogation. There are a number of legitimate reasons why a mode-S transponder will not reply, and a second interrogation is made to attempt

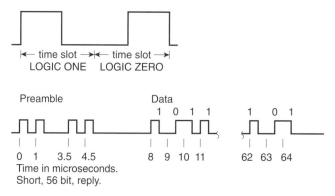

Figure 7-9 Format of a mode-S reply.

to establish a two-way communication. Some of these reasons are that the transponder is replying to another interrogator, two interrogations are garbled, or the transponder is temporarily suppressed by other L-band equipment aboard the aircraft.

COLLISION AVOIDANCE

Although midair collisions are rare, when they do occur, they involve considerable loss of life both in the aircraft involved and potentially on the ground. Government and airline pilots groups have recommended the development of a collision avoidance system for a number of years. A number of systems were developed and demonstrated, but the systems required very advanced and expensive technology. Finally, in the 1980s a collision avoidance system called TCAS, for traffic and collision avoidance system, was developed and installed in air transport aircraft.

TCAS uses the ATCRBS or mode-S transponders already installed aboard aircraft for gathering information relative to the position and altitude of aircraft. This information can be monitored to determine if any nearby aircraft presents a threat of collision. But just what constitutes the threat of collision?

To quantify the threat of collision, the concept of tau was developed by early researchers into the mechanics of midair collisions. *Tau* is the range of an aircraft divided by the range rate. *Range* is the distance from the protected aircraft, called *own* aircraft, to another aircraft. Range is not a vector, but the distance from own aircraft to another aircraft regardless of whether the aircraft is directly ahead, to the side, above, or below.

The rate at which the range changes as a function of time is the *range rate*. If the aircraft are separating, the range becomes larger and the range rate is positive. If the aircraft are coming closer, the range decreases and the range rate is negative.

Range divided by range rate produces a quantity with the dimension of time. This is tau, which is essentially the amount of time, in tau seconds, that the range will be zero and a collision will occur, if the range continues to decrease at the rate it was when tau was calculated.

Of course, it is not certain that a collision will occur in tau seconds; tau implies that the aircraft are approaching, and if they continue to approach at that rate, they will collide in tau seconds. It is possible that the aircraft are not on a collision course at all. It is also possible that either aircraft could change its direction of flight such that the range increases and thus there is no danger of collision. It is not possible for any collision avoidance system to predict the future, so a value of tau that is, essentially, too close for comfort is chosen, and the collision avoidance system will advise the aircrew of a potential hazard.

It is helpful to have an understanding of the protected space around an aircraft. The most important thing to remember is that the threat of a collision is not a function of distance but of tau. Therefore, it is not possible to construct a space around an aircraft and to sound an alarm whenever an aircraft has entered this protected airspace. As an example, two aircraft passing at two different cruising altitudes can pass safely within a few thousand feet. Two aircraft at the same altitude and passing within a few thousand feet would be a different matter.

Photo 7–2   The complete TCAs equipment showing antennas, IVSIs, control heads, and the LRUs. Photo courtesy of Allied Signal Commercial Avionics Systems.

The protected airspace does have a generalized shape. First, aircraft forward of the protected aircraft involved the greatest distances. Obviously, the range rate will be the greatest in front of the aircraft because the aircraft's forward velocity will add to the velocity of the other aircraft and will require monitoring of aircraft at greater distances. To the rear of the aircraft is the opposite situation. Now, the forward velocity of the protected aircraft subtracts from the forward velocity of the other aircraft.

It should not be forgotten that collision avoidance is a three-dimensional problem and that aircraft above and below should be monitored. The vertical speed of an aircraft is typically less than 1 nmi/minute or 60 knots. Therefore, only a small volume of airspace would be protected above and below the aircraft. The result is shown in Fig. 7-10, a flattened tear-drop shape with the majority of the protected airspace in front of the aircraft.

TCAS monitors the tau of nearby aircraft and sounds an alarm if the value of tau becomes less than a preset number. The first level of warning is called a *traffic advisory,* or a TA, and occurs when an aircraft has a tau value of 40 s.

Most TCAS systems have a position display showing those aircraft being tracked by the TCAS system, as shown in Fig. 7-11. Aircraft that have a tau of greater than 40 s are called *proximate* aircraft. Once an aircraft has a tau value of 40 s, the appearance of the aircraft's display changes and a voice warning is activated: "traffic, traffic."

Of all the aircraft displayed on the TCAS display, a traffic advisory does not necessarily issue from the aircraft shown as the closest. The TCAS display shows physical separation, but it is tau that issues a warning. Aircraft that are close and not getting closer do not pose a threat. Displayed aircraft that are very close could be several thousand feet above or below own aircraft and not produce a threat. At the issuance of the TA, the aircrew should locate the intruder on the TCAS display and then visually, if possible, and prepare for an evasive maneuver.

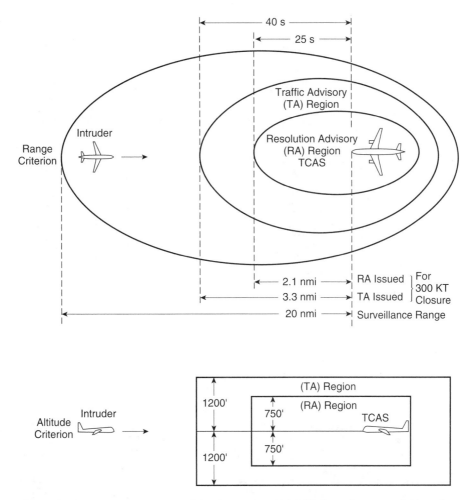

Figure 7–10   Approximate shape of the airspace in which TCAS monitors potential intruders.

At tau equal to 25 s, the TCAS system will issue a resolution advisory, or RA. At tau equal to 25 s there is not enough time to locate the intruder and prepare an evasive maneuver, and thus the TCAS system provides the evasive maneuver.

The most common TCAS system, TCAS II, provides only vertical evasive maneuvers of which there are two broad categories, corrective and preventive. The corrective maneuver requires that the aircrew alter its vertical speed. As an example, if the aircraft is climbing, a possible corrective maneuver would be to stop climbing or to descend. Another example is an aircraft in level flight for which the evasive corrective maneuver would be to descend. A preventive maneuver would be for an aircraft not to change its current vertical speed.

WHAT HAPPENS DURING A TCAS ENCOUNTER

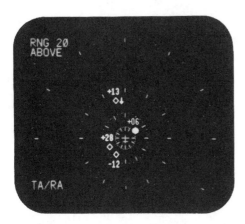

1. The Bendix/King ATAD TCAS display showing Mode C targets. The +13 and the ↓ on the upper target indicates that aircraft is 1300 feet above your aircraft and descending at 500 feet per minute or greater. A Traffic Advisory target is indicated 2 miles at 2 o'clock by the round symbol.

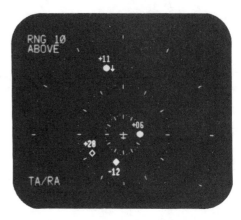

2. Two intruders have now become Traffic Advisory targets; the target at 3 miles and 6:30 position has become a "Proximity Target," meaning it is within 6 miles and ±1200 feet of your altitude. Proximity Targets are displayed as solid diamonds only when there is another Traffic or Resolution Advisory in process.

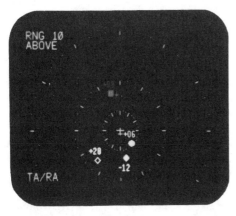

3. The target at 4 miles and 11:30 has become a Resolution Advisory. The pilot will find the actual vertical rate requirements on the VSI.

4. The Resolution Advisory has been downgraded to a Traffic Advisory and the "encounter" is over.

Figure 7–11 Example of a TCAS position display. Photo courtesy of Allied Signal Commercial Avionics Systems.

An RA, like a TA, is initiated with a voice announcement. As an example, for an aircraft in level flight for which the evasive maneuver is to climb, the voice announcement is "climb, climb, climb." For a preventative maneuver, the voice message is "monitor vertical speed."

The allowed vertical speeds are indicated by the illumination of green and red bands on the vertical speed indicator, or VSI. Figure 7-12 shows example voice announcements and corresponding VSI indications.

How is it that, when a collision is predicted, the evasive maneuver is to not deviate from current vertical speed? This would occur when two TCAS-equipped aircraft were on a collision course and one aircraft issued an RA. The aircraft issuing the RA transmits it to the intruder aircraft over the mode-S link, which inhibits the issuance of its own RA. If both TCAS-equipped aircraft are tracking each other and should calculate roughly the same value of tau, they would then issue an RA nearly at the same time. To prevent the issuance of a second RA that nullifies the first RA, a system of coordination is used. An example would be two aircraft that are approaching head on at the same altitude, and the first aircraft issues a "climb" RA, which would effectively avoid a collision. The second aircraft, seeing the same collision scenario, would shortly after the first RA issue a second RA, which would most likely be for the same maneuver, climb, which would nullify the first evasive maneuver.

To prevent this, the first aircraft broadcasts the fact that it issued an RA, and the second aircraft, receiving the broadcast, issues a preventive RA, "don't climb." If, by chance, both aircraft issued an RA simultaneously, the aircraft with the higher mode-S address is given the evasive maneuver and the second aircraft cancels its corrective RA and issues a preventive RA instead.

It is clear that in some situations not all RAs can be enacted. As an example, if an aircraft is approaching an airport for a landing and a descend RA is issued, the aircraft cannot take advantage of the RA. Therefore, the TCAS system is adjusted as a function of the aircraft's altitude, AGL, to prevent certain RAs from being issued. This adjustment is called *sensitivity level,* as shown in Fig. 7-13.

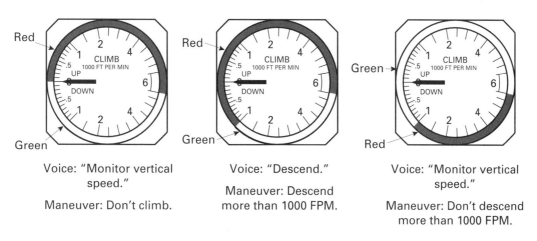

Figure 7–12   Example RAs showing the VSI indicators and the voice announcements.

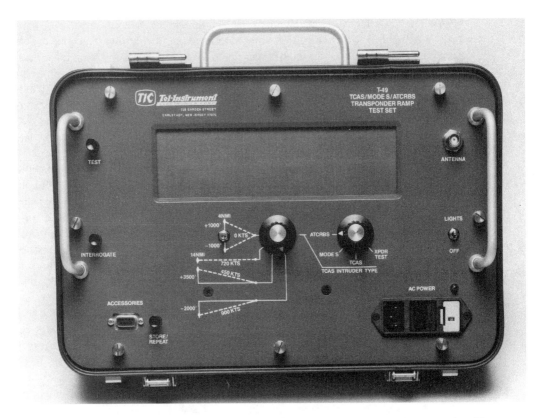

Photo 7–3   The Tel-Instrument T-49 is a ramp test set for TCAs and transponders. Photo courtesy of Tel-Instrument Electronics Corp.

Altitude	Sensitivity level	Tau Values for Advisories	
		TA	RA
0-500′ AGL	2	20	Not issued
500-2500′ AGL	4	35	20
2500-10,000 MSL	5	40	25
10,000 – 20,000 MSL	6	45	30
Above 20,000 MSL	7	48	35

Figure 7–13   Sensitivity levels as a function of aircraft altitude.

The highest sensitivity level occurs at high altitudes where evasive maneuvers may be made without danger and where aircraft should be widely separated. Therefore, TAs and RAs are issued for the largest values. The other end of the spectrum is for low altitudes, AGL, where aircraft are typically close and the ability to perform evasive maneuvers is diminished. These represent the lower sensitivity levels for which smaller values of tau are employed.

It is the task of TCAS to determine the existence and position of aircraft within 10 or so nautical miles and to monitor these aircraft to determine which may pose a threat to own aircraft. The TCAS system operates totally through communications with secondary radar transponders, both ATCRBS and mode-S. As with any secondary radar system, the range of the intruder aircraft is determined by measuring the elapsed time from an interrogation to the receipt of a reply. Altitude is determined from either an ATCRBS mode-C reply or from a mode-S surveillance-altitude reply. The angular position of the intruder is determined by TCAS by the use of a directional antenna.

TCAS is an airborne interrogator and suffers all the problems of any interrogator, such as multipath and garble. When TCAS has a mode-S equipped aircraft under surveillance, there is no garble because of the nature of the mode-S system. However, ATCRBS transponders offer no protection from garble, and innovative techniques are used with TCAS systems to reduce the amount of garble.

To reduce the amount of garble from ATCRBS transponders, TCAS transmits interrogations of variable power levels. Also, additional pulses are transmitted 2 μs ahead of an ATCRBS interrogation, which, if received, would cause the transponder to suppress. These additional pulses are also transmitted at various power levels. If two transponders garble, of the various combinations of power levels involved in the interrogation, usually one transponder is suppressed while the second replies, and vice versa. This technique is called *whisper–shout,* which alludes to the strong and weak interrogations employed.

Once an ungarbled reply is available from a transponder, this transponder is placed on a *roll call* and is tracked. As long as the aircraft is within the operating range of the TCAS display, which is between 10 and 20 nmi, the transponder is tracked by TCAS and displayed. Once the transponder has a range greater than the operational range of the TCAS, it is dropped and is no longer tracked.

TCAS I and TCAS III

TCAS II is the system found on all air transport aircraft at the time of the writing of this text. An enhanced TCAS system that allows for full three-dimensional RAs, called TCAS III, is expected in the future. Aircraft are expected to upgrade from TCAS II to TCAS III by the installation of new software.

A simpler TCAS that will be installed in general aviation and smaller commuter airliners is the TCAS I system. There are two permitted methods of implementing TCAS I: active and passive. One method of implementing passive TCAS I is to monitor the signal strength of nearby transponders. Unfortunately, this method has been shown to be unreliable because of the large variations in transmitter power outputs, antenna gains, and so on.

The passive TCAS I system can rely on signal strength or can monitor interrogations and replies without regard to signal strength and, by mathematically manipulating the time differences between the interrogations and replies, along with information relative to the location of interrogators and the reported altitude, can determine the position of nearby aircraft. The major attraction of passive TCAS I is that it does not require a complex transmitter capable of transmitting the whisper–shout sequence of transmissions. However, it does require an enormous data base for the location of ground interrogators and considerable computation time. This technique has been abandoned in favor of the active TCAS I system that interrogates nearby transponders much like TCAS II.

The major difference between a TCAS I system and its more sophisticated relative, TCAS II, is that the TCAS I system cannot issue an RA.

WEATHER RADAR

The surveillance systems discussed so far are used to monitor the location of aircraft either for collision avoidance or air traffic control. Weather radar is used to monitor the location of weather formations. *Weather radar* is an airborne primary radar system operating in the X band at 9375 MHz.

Like ground-based radars, weather radar emits a high-power narrow pulse of RF energy that is backscattered from the weather formations and received by the weather radar receiver. What has made the design of weather radar difficult is to create a radar system that is small and lightweight.

The signal returns from weather radar are a function of the amount of moisture in the air. Waterdrops or rain, as opposed to water vapor or clouds, provides much stronger signal returns. The greater the rain rate or the more intense the storm is, the greater the strength of the return signals.

The block diagram of the weather radar is shown in Fig. 7-14. The transmitter is a high-powered vacuum-tube oscillator called a *magnetron*. The transmitter feeds the antenna through a transmit–receive (T–R) switch, which allows the antenna to be shared between the receiver and transmitter. The weather radar receiver is a superheterodyne receiver. The weather radar receiver employs one of two types of IF amplifiers: the logarithmic or log amplifier and a sensitivity time control, or STC.

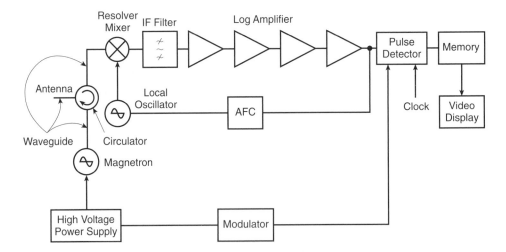

Figure 7–14 Block diagram of a weather radar system.

The signal strength is a function of the density of moisture in a weather formation and the range to the target. Thus, the return signals received shortly after the transmit pulse are the least attenuated by distance. Later signals are much weaker, primarily due to the attenuation of distance. The attenuation of radar signals as a function of range is proportional to the negative fourth power of the range. As an example, assume that a weather radar is to operate with storm formations as close as 5 nmi and as far as 200 nmi. The signals from 10 nmi distant are 1/16th of those received from 5 nmi for the same-intensity storm. Those from 20 nmi are 1/256th of those from 5 nmi. It would make sense to increase the sensitivity of the receiver as a function of time from the transmitted pulse to accommodate the decrease of signal because of the longer distances traveled.

Example

How is the receiver gain changed as a function of time to achieve a demodulated return signal that is corrected for attenuation due to distance?

Solution.   The gain should increase as the distance, R, to the fourth power to compensate for a signal reduction of $1/R^4$. This distance is $R = ct$, where R is the distance, c the velocity of light, and t the elapsed time from the transmission of the pulse. Therefore, the gain should be

$$\text{gain} = G(ct)^4$$

where G is the gain required at $ct = 1$ meter (m).

Another approach to the solution is that, every time the distance is doubled, the signal power is reduced by a factor of $1/2^4$, or 12 dB. This implies that the gain must be increased by 12 dB for each doubling of the signal.

$$\text{gain (dB)} = G_0 + 40 \log \frac{ct}{1 \text{ nmi}}$$

where G_0 is the gain required for return signals from 1 nmi.

This equation could also be derived by taking $10 \times \log$ or the decibel value of both sides of the previous equation.

Example

How much receiver gain is required for a weather radar receiver to receive returns from 200 nmi if 80 dB of gain is required to receive returns from 5 nmi?

Solution.   Because the gain at 5 nmi is given, rather than the gain at 1 nmi, the equivalent 1-nmi gain must be calculated. Because the signal strength is inversely proportional to the distance to the fourth power, the gain required for signals arriving from 1 nmi is 5 to the fourth power, or 1/625th of that required for signals from 5 nmi. The 80-dB gain required at 5 nmi is reduced by $10 \log (625)$, or 28 dB. Thus the gain required for signals from 1 nmi is 52 dB. The gain is

$$\text{gain (dB)} = 52 \text{ dB} + 40 \log \frac{ct}{1 \text{ nmi}} = 52 \text{ dB} + 40 \log (200)$$

$$= 52 \text{ dB} + 92 \text{ dB} = 144 \text{ dB}.$$

Many weather radar transmitters use a high-power microwave oscillator, the *mag-netron,* for the transmitter. This vacuum tube was developed during World War II and was the major source of high-power microwave energy for many years. It is still used in radar equipment, including weather radar. The magnetron employs resonant cavities placed in a magnetic field that contains a heated cathode and a vacuum. A high voltage is employed that causes a stream of electrons to be emitted and, because of the magnetic field, the electrons are deflected in a circular path. This path causes energy at microwave frequencies to be generated in the resonant cavities. Because the cavities can support oscillation at only one frequency, the energy is added at that frequency. Actually, the resonant cavities can support oscillations at odd multiples of their resonant frequency. To prevent oscillations at a harmonic, called *moding,* which means that the magnetron operates at different modes of oscillations the cavities are modified by a technique called *strapping.*

The magnetron is an effective and relatively low cost method of generating high-power microwave energy. The magnetron's frequency tends to wander, and the weather radar receiver must either track the wandering magnetron or employ a bandwidth sufficient to encompass the greatest frequency variation. Employing more bandwidth than is necessary is undesirable because greater receiver bandwidth allows more noise to be admitted and reduces the sensitivity of the receiver. If a transmitter with superior frequency control

Photo 7–4   A weather radar installed in an aircraft instrument panel. Photo courtesy of Rockwell International Corp., Collins Divisions.

is employed, a narrowband receiver can be used to provide enhanced receiver sensitivity. A more sensitive receiver implies that a lower-power transmitter can be used for the same radar range.

The current trend in the design of weather radar transmitters is the use of a crystal-controlled transmitter employing lower-power output and a significantly narrower receiver. The receiver local oscillator is derived from the transmitter frequency controlling element, and thus the receiver local oscillator is phase locked or *coherent* with the transmitter oscillator. This type of weather radar offers two advantages. First, because the transmitter frequency is controlled with a quartz crystal, the receiver bandwidth can be as narrow as necessary to pass the received signals. This enhances receiver sensitivity.

A second important characteristic is that the Doppler shift of the returned signals may be measured. Unlike the magnetron, the quartz crystal frequency reference is functioning continuously even during periods when the transmitter is not transmitting. This provides a frequency reference against which the return signals may be compared to assess the amount of frequency shift that may occur on the return signals. Although the receiver in a magnetron type of radar tracks the transmitter, the magnetron is so unstable that it would be nearly impossible to determine any frequency shift due to Doppler shift with a magnetron transmitter.

The Doppler shift is produced by the turbulence of the waterdrops in the storm system. This provides the radar with information relative to the turbulence, as well as to the intensity of the rain in a storm cell. These weather radar systems are called either coherent or turbulence mode weather radars and are very effective in mapping dangerous weather.

Another technique of variable-width transmitter pulses is used primarily with coherent weather radars. The resolution of a radar is a function of the pulse width used. A pulse of 6.2 μs is 1 mi long as it propagates from transmitter back to the receiver. To achieve a radar resolution of better than 1, nmi pulse widths of less than 6.2 μs are required. Weather radars using magnetrons typically use between 0.5- and 2-μs pulses.

The amount of energy in a pulse, which affects the range of a radar, is directly proportional to the pulse width. Also, longer pulses have a more narrow spectrum, and a narrower receiver bandwidth may be employed. Fortunately, when resolution is desired, for nearby weather a narrow pulse may be used. For distant weather, extreme resolution is not required. To enhance the range of a radar, short pulses are used for mapping weather that is nearby, whereas longer pulses are used for mapping weather at a distance. Two receiver bandwidths are used. A wide-bandwidth receiver is used for the narrow-width nearby returns, and a narrowband receiver with a lower noise level and greater sensitivity is used for the longer-distance returns.

LOW-FREQUENCY WEATHER MAPPING

Thunderstorms are perhaps the most dangerous type of weather activity. This is because they tend to appear suddenly, without warning, and quickly become dangerous storms. In addition to high-altitude clouds and large amounts of precipitation, thunderstorms produce electrical discharges that produce radio static.

For many years, pilots have observed that precipitation static caused ADF receivers to point to the source of the lightning strikes. This was due to the fact that lightning strikes generate strong broadband radiation that is picked up by an ADF receiver and processed like any other signal, which causes the ADF to point to the source of lightning. The strength of the static heard in the ADF gives an indication of the distance to the lightning discharge.

To use signal strength as a measure of range when plotting lightning strikes, it is necessary to assume that all strikes contain roughly the same energy. This turns out to be true, and thus distance can be estimated by the received strength of the lightning strike.

The low-frequency weather mapping system uses a circuit very much like an ADF, but stores the range and direction of the lightning strikes. The block diagram of the low-frequency weather mapping system is shown in Fig. 7-15. As with the ADF, both sense and loop antennas are used. The loop antennas are cross-loop antennas wound on a ferrite block. The low-frequency weather mapping system is tuned to one frequency, around 50 kilohertz (kHz). There is no reason to employ a superheterodyne receiver, and signal processing is accomplished by using active filters and operational amplifiers.

The electric field signal is used as a reference to demodulate the outputs from the crossed-loop antennas. The electric field signal is amplified, phase shifted by 90°, and hard limited. This output is used as one input to two balanced mixers. The other input to the two balanced mixers is the amplified and filtered outputs of the two loops. The outputs of the balanced mixers are dc levels that are proportional to the amplitude of the loop outputs, which are the X and Y components of the signal strength.

Both the weather radar and the low-frequency weather mapping systems have advantages and disadvantages. The major difference between the two systems is that the low-frequency weather mapping system is much less expensive than the weather radar. Generally, the weather radar is a more accurate and precise measuring tool for locating severe

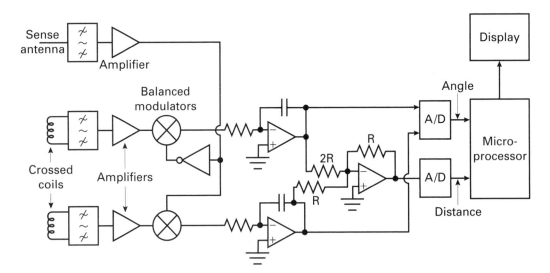

Figure 7–15   Block diagram of a low-frequency weather mapping system.

weather. However, there are some significant advantages of the low-frequency weather mapping system.

First, some storms have severe lightning and turbulence without heavy rain. Also, heavy rain does not always imply a severe storm. The newer turbulence Doppler weather radars can provide information about the wind conditions and thus a more accurate picture of weather. Second, the low-frequency weather mapping system can display weather on the ground, whereas weather radar cannot. Finally, the low-frequency weather mapping system can provide a 360° picture of weather.

The ideal weather mapping tool would be a combination of low-frequency weather mapping and coherent weather radar. This would provide a weather picture that includes rainfall rate, turbulence, and lightning strikes.

REVIEW QUESTIONS

7.1. What is a secondary radar system?

7.2. What benefits are available from a secondary radar system?

7.3. How many different squawks are available from an ATCRBS transponder?

7.4. What is the transponder emergency code?

7.5. Why is the Gray code used for transmitting altitude information?

7.6. What is fruit? How is fruit eliminated?

7.7. What is garble? What techniques are used to eliminate garble?

7.8. How much elapsed time passes from the leading edge of a mode-A interrogation and the receipt of the leading edge of the first framing pulse of a reply for a transponder 50 nmi from the interrogator?

7.9. What methods are employed to eliminate garble?

7.10. On what frequency does the airborne transponder receive interrogations?

7.11. On what frequency does the airborne transponder reply?

7.12. What type of warnings are available from a TCAS II-equipped aircraft relative to another TCAS II-equipped aircraft? Or to an ATCRBS mode-C-equipped aircraft?

7.13. What types of maneuvers are available from a TCAS II system?

7.14. How do preventive and corrective maneuvers differ?

7.15. What types of maneuvers will the TCAS III system be able to advise?

8

Instruments and Displays

CHAPTER OBJECTIVES

After completing this chapter the student will have an understanding of basic aircraft instruments, particularly those that combine radio and gyroscopic information. Also, display technologies are discussed.

INTRODUCTION

The VOR indicator and the ADF indicator have been discussed previously. The VOR indicator provided navigation information relative to a selected course. Although the aircraft heading has no effect on the displayed information, if the aircraft is flown in the "wrong" direction, that is, away from a VOR on a TO omni bearing or toward the VOR on a FROM bearing, the sense of the indicator pointer is reversed. The ADF indicator involves the aircraft heading, in that the indicator literally points to the ADF. Therefore, the ADF indicator turns with the aircraft.

Almost all navigation is relative to magnetic north, which makes the magnetic compass an important instrument for setting the aircraft heading. Compasses have some difficulties when used directly as a navigation instrument. One major problem is that the compass produces an error when the aircraft is in a turn. A second problem is that the compass tends to respond slowly and also oscillates around the actual heading. This is particularly true when the aircraft is turning or in rough air.

The development of navigation instruments has been toward combining indicators that provide stable indications in a format that is useful to the pilot. Data that are used together are put into one combination indicator to ease the task of navigation and to provide a better mental picture of the navigation situation. An example would be the combination of heading and VOR information in one indicator.

THE GYROSCOPE

A more stable directional reference is used in the aircraft to give a more steady, error-free heading indication based on the stability of a spinning wheel. A spinning wheel resists turning. All objects resist being turned because of their inertia. But, if a well-balanced wheel is rotated at a high rate, its resistance to being turned will increase well beyond its normal inertia. This is the principle behind a toy top that seems to resist gravity by remaining upright and spinning. The spinning wheel used for providing stable inertial references is called a *gyroscope*.

The toy top is an example of a gyro with a vertical spin axis that tends to remain vertical. There are two aircraft axis that may be referenced to the vertical gyro, roll and pitch. When the gyro spin axis is horizontal, the gyro tends to remain aligned in one direction, such as pointing toward north. This type of gyro orientation is used in directional gyros.

To provide an indicator of direction, the gyroscope is mounted in a structure that allows the gyro to remain stationary as the aircraft moves around it. Such a structure is shown in Fig. 8-1. This mounting structure is called a *gimbal*.

The gimbal assembly for a directional gyro must allow the aircraft to roll and pitch without affecting the gyro. This requires two pivoted assemblies, one to allow roll and the other to allow pitch without forcing the gyro to move. The pivot along the roll axis allows the aircraft to roll without affecting the gyro, while the gyro's own shaft aligned along the pitch axis allows the aircraft to pitch without affecting the gyro.

The pivot along the yaw axis, which is connected to the outer gimbal, allows the aircraft to yaw without affecting the gyro. This permits the aircraft to turn while the gyro remains aligned to its preset direction. By connecting a compass indicator to the gyro wheel, the gyro is used to provide the heading angle.

Most gyro instruments require a flat face indicator and, rather than the indicator being a moving scale as shown in Fig. 8-1, an angle drive is used connected to the outer gimbal to drive a vertical flat face indicator. The moving scale indicator shown in the figure was used for many years, however.

The gyro spin axis may be oriented in any direction as long as the indicator provides a directional reference that points to magnetic north. A gyro indicator is arranged so that the relationship between the spin axis and the indicator may be adjusted. Before a flight, while the aircraft is on the ground with no turbulence or turning errors, the gyro is set to match the magnetic compass by the use of a front-panel knob. The gyro is set by releasing the indicator from the gyro and turning the indicator to match the magnetic compass. Wherever the spin axis of the gyro happens to be, it remains there and has no specific relationship to magnetic north.

In the previous discussion of the gimbal assembly, the X axis was the roll axis of the aircraft and the Y axis was the pitch. If the aircraft is turned, this is no longer the case. The

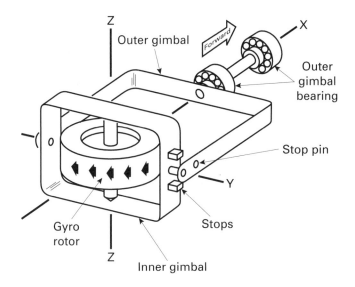

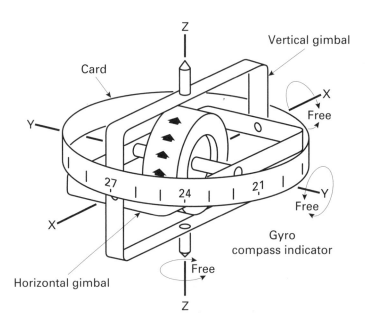

Figure 8–1 Vertical and directional gyroscope with mounting structures or gimbals.

Y axis could be the roll axis while the X axis becomes the pitch. Most likely, neither the X or Y axis would be pitch or roll. The gimbal will still function to isolate the gyro from the pitch and roll movements of the aircraft.

Some orientations of the gimbals prevent the necessary degrees of freedom required of a gyro. Such orientation is called *gimbal lock* and should be avoided. Usually, the gimbals are configured so that it is nearly impossible to place a gyro in a gimbal lock situation. Typically, the angles involved are for pitch and roll angles of more than 45°. If a gyro is placed in a gimbal lock condition, it can be forced from its normal position; when this happens, the gyro is said to have tumbled. Aircraft performing aerobatic maneuvers may be able to tumble a gyro, but most aircraft of the normal and utility categories are not capable of tumbling a gyro.

Extraneous forces applied to the gyro wheel due to frictions within the gyro will cause the gyro to move slightly from its set position. This movement is called *precession,* or *gyro drift.* Also, the gyro is affected by the acceleration due to the rotation of Earth. Even a gyro that is stationary is actually moving in accelerated, circular, motion at nearly 1000 knots (KTs). Because of this, the gyro must be periodically reset to the correct magnetic heading.

Gyros are used to provide more than just directional reference. The gyro can provide an artificial horizon for pitch guidance and rate-of-turn information. Figure 8-1 shows the gimbal arrangement for an artificial horizon. As with the directional gyro, the gimbal assembly allows the gyro complete freedom of motion except that the spin axis is vertical. The X axis is the roll axis while the Y axis is the pitch axis. Unlike the directional gyro, the X and Y axes are forced to turn with the aircraft and will always be the roll and pitch axis. Therefore, the gyro will remain vertical and will provide an indication of roll and pitch angles. The familiar blue and black disk simulating the sky and ground is attached to the gyro and provides a two-dimensional reference to the horizon for pitch and roll. The first application of the vertical gyro was the artificial horizons used for the very first blind flight in the 1920s.

The vertical gyro does not need an erecting system because the gyro is made to have a small offset in the center of gravity of the indicator, which causes the gyro to assume a vertical position when the aircraft is on the ground and the gyro is not powered. In larger aircraft a vertical gyro can be used to provide pitch and roll information for other systems such as the autopilot.

The major attitude-indicating device in larger aircraft is the attitude direction indicator, or ADI. This indicator is very similar to the artificial horizon but includes indicators associated with the ILS system for aid in landing.

The gyro wheel is set in motion by either an electric motor or an air motor. The more mass the gyro wheel has, the better the performance of the gyro will be. This is also true of the rotational speed of the gyro; the faster the gyro is turning, the better the gyro will resist drifting. The electric or air motor is not a part of the gyro wheel, and to maximize the spinning mass while minimizing the "dead weight," only a small motor is used to power the gyro wheel. The dead weight is those parts of the gyro that must be suspended in the gimbal but do not constitute a part of the spinning gyro mass. When the gyro has come to full operating speed, the motor needs to have only enough torque to overcome the friction of the bearings. The friction of the bearings causes gyro drift, and to reduce the amount of drift, only very low friction bearings are used and thus only a small motor is required to keep the gyro spinning.

The motor must, however, bring the gyro wheel up to speed from stationary and this can take several minutes. Typical gyro rotational speeds are from 10,000 to 20,000 rpm. During the time the gyro is being accelerated to its operating speed, the gyro cannot be adjusted to magnetic north because the amount of drift during this period can be great. This period of time is often called the *spin up* of a gyro.

Because of the large mass and low friction bearings, it can take up to 15 minutes for a gyro to stop spinning. During this period of time, the gyro is vulnerable to damage if the instrument is moved. Therefore, at least 15 min should elapse from when a gyro is de-energized and it is removed from an aircraft.

Gyros are powered either by air or electricity. Most aircraft have both types of gyros on board so that, if either the supply of electricity or air fails, the remaining gyro can derive its power from the alternative source.

Figure 8-2 shows an air-driven gyro. The outer periphery of the gyro wheel has indentations called *buckets* that catch the air from an airstream and cause the wheel to spin. A source of powering air must be provided to the gyro wheel through the gimbal system. The gimbals are hollow and are interconnected with air joints that allow for mechanical rotation but pass air.

Air-driven gyros are sometimes called vacuum gyros, because the gyro is mounted in an airtight case and the case is connected to a source of vacuum. Air is drawn into the case because of the vacuum, and the entering air is directed at the buckets of the gyro to cause it to spin. The net result is the same; compressed air flows over the buckets and causes the gyro to spin.

Electric gyros have an electric motor mounted on the gyro wheel to overcome the friction of the bearings. A source of electrical current must be made available to the motor, and this must be done through the rotating gimbal joints. Rotating electrical contacts called *slip rings* are provided for a path of the electrical current.

When the gyro is not spinning, the gyro orientation can be any relationship and must be forced to be erect. In the case of a directional gyro, the axis of the spinning wheel must be horizontal. As explained previously, the orientation of the axis of the gyro around the vertical axis is not critical because the gyro will be set to match the magnetic compass. However, the requirement for a directional gyro that the spin axis be horizontal is critical.

The gyro can be set erect in several ways. First, the gyro can be caged, which means that a mechanical system is enabled that forces the gyro to a specific position. This mechanical erecting system completely surrounds the gyro somewhat like a cage and forces the gimbals to assume specific positions. This was usually done in older, inexpensive panel-mounted gyros by pulling a knob on the panel of the instrument, a maneuver called *caging* the gyro.

When a gyro is inaccessible to the front panel, an automatic system is employed called an *automatic erecting system*. Even when a manual caging system is provided, it is better not to rely on a pilot to remember to cage the gyro, and an automatic erecting system is employed. The automatic erecting system employed in a gyro, matches the type of gyro.

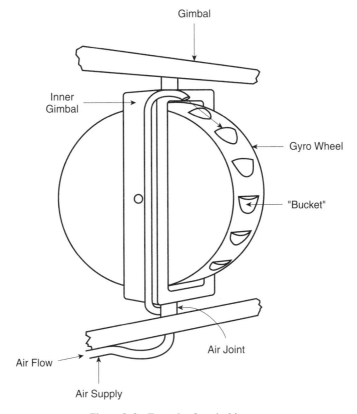

Figure 8–2   Example of an air-driven gyro.

That is, an electrically operated system is used for electric gyros and an air-operated system is used for air-operated gyros.

An example of each type of automatic erecting system is shown in Fig. 8-3. The erecting system will orient the gyro in its normal operating position when the gyro is spun up. This usually occurs when the aircraft is on the ground. Therefore, the aircraft may be considered to be horizontal. This figure bears no resemblance to the thin gimbal rings shown in Fig. 8-1 or the wheel with buckets shown in Fig. 8-2. These figures were simplified to aid in understanding the principles involved. Practical gyros use much more substantial, but still lightweight, gimbal structures.

The air-operated automatic erecting system relies on two pendulums that provide a downward reference. If the gyro is tilted, one pendulum opens an air passage and causes a jet of air to push the gyro to the erect position. Once the gyro has reached an erect position, both air gates are closed off and the gyro remains in the erect position.

The electric erecting system senses the position of the gimbals using mercury switches and moves the gyro using an electromagnet arrangement called a *torquer.* The example shown in Fig. 8-3 uses two coils of wire and a magnet attached to the outer gimbal.

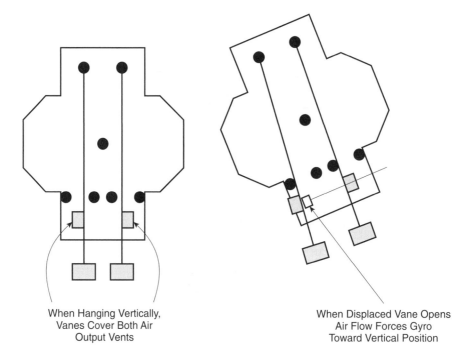

When Hanging Vertically,
Vanes Cover Both Air
Output Vents

When Displaced Vane Opens
Air Flow Forces Gyro
Toward Vertical Position

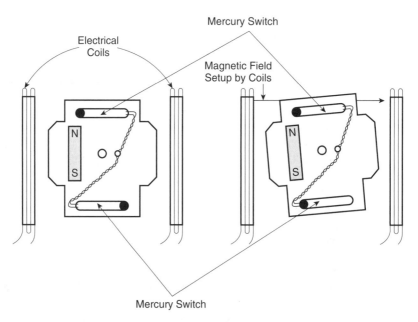

Figure 8–3   Examples of air and electric automatic erecting systems.

Photo 8–1   This slaved directional gyro can be referenced to either magnetic or true
north. This unit is manufactured to full military specifications. Photo courtesy of
Aeronetics, a division of AAR Manufacturing, Inc.

RATE GYRO

Although gyros resist being turned, when they are forced to turn, the gyro produces a resisting force that is directed 90° from the applied force, as shown in Fig. 8-4.

The directional gyro is mounted freely so that it will not turn and thus provides the
directional reference. However, if a gyro is forced to turn, the amount of resisting force
resulting from the turning of a gyro is proportional to the rate of turn.

To measure the rate of turn, the gyro wheel is mounted horizontally, as a directional
gyro, except the gyro is forced to turn with the aircraft. When the aircraft turns, the gyro
will rotate to the left or right depending on the direction of the turn, and the force will work
against a spring. The amount that the spring deflects is displayed on an indicator and is
proportional to the rate of turn, as shown in Fig. 8-5.

GYRO-BASED NAVIGATION INSTRUMENTS

Radio and heading instruments are often used together in navigation, and a number of instruments combine radio navigation and heading information. One of the earliest of such
instruments is the *radio magnetic indicator,* or RMI.

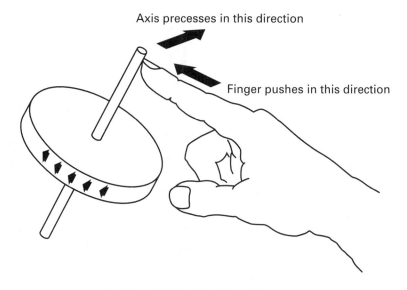

Figure 8–4   Resisting forces of a gyro.

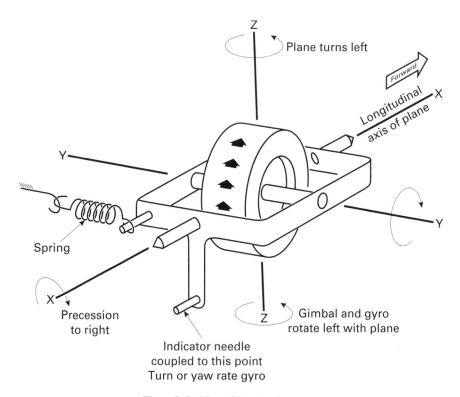

Figure 8–5   View of the rate-of-turn gyro.

The RMI combines a gyroscopic heading indicator and either an ADF pointer or a VOR indicator, as shown in Fig. 8-6. The VOR indicator, rather than having a course-deviation indicator and an OBS, points to the VOR much like an ADF points to the NDB. The purpose of the RMI is to aid the pilot in visualizing the location of the NDB or the VOR relative to north. This is particularly handy when a heading must be flown to make an approach for landing. To further enhance the visual picture, some RMIs have two pointers, one each for the ADF and VOR.

The RMI in its basic form, the ADF RMI, is a simple application of two pointing-type indicators in one instrument. Essentially, the two instruments are mounted in the same case and share the same shaft. The heading gyro points to north while the ADF points to the NDB. There is no interaction between the ADF and the directional gyro.

When the VOR is used in an RMI, there must be interaction between the directional gyro and the VOR. Remember, a normal VOR indicator has no relationship to the heading of the aircraft. In the RMI, the VOR pointer will physically point to the VOR that requires that the heading of the aircraft be involved in the display of the VOR information. A special VOR converter is used that provides an indicator of the VOR radial, rather than a CDI and TO/FROM indication. The VOR pointer should point to the VOR radial on the compass card that rotates relative to the heading of the aircraft. There are several methods of achieving this relationship. First, the pointing device is mounted on the compass card so that the VOR pointer moves with the compass card. A second approach is to subtract the aircraft heading from the VOR radial and display the difference. The advantage of the sec-

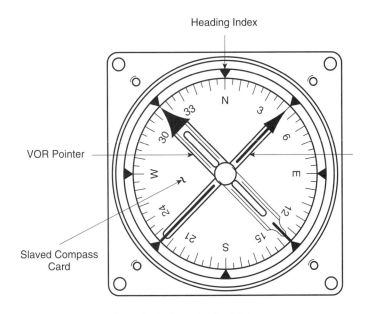

Figure 8–6   Example of an RMI.

ond method is that the VOR indicator does not have to be rotated and can remain stationary. Both methods are used.

A more sophisticated instrument that combines the gyro and the VOR with a display that gives one of the best visualizations available is the *horizontal situation indicator,* or HSI, as shown in Fig. 8-7. This indicator is an extension of the VOR–RMI and uses a course deviation indicator and an OBS, which allows the VOR to be used in a more conventional fashion. To enhance the pictorial nature of this instrument, an outline drawing, or an icon, of an aircraft is at the center of the instrument.

The major difference between the HSI and the VOR–RMI, is that the desired VOR radial is selected using an OBS. The HSI displays the aircraft's position relative to a selected radial. The VOR–RMI has no course selection and shows the aircraft's situation without reference to a selected bearing. In the HSI, the VOR pointer is an arrow with a movable center section that serves as a CDI. Unlike the VOR–RMI, the arrow points to the selected VOR radial on the compass card. Like the RMI, the compass card rotates as a function of heading, and the VOR arrow rotates with the compass card. The head of the VOR arrow is the TO/FROM indicator and shows where the VOR is located relative to the selected radial. The center section shows the deviation from the selected radial and gives a picture of the intercept angle of the aircraft to the selected radial.

The rotating compass card of the HSI contains the OBS and the CDI. A structure as complex as this is not driven directly from a directional gyro because of the amount of friction involved; it requires the assistance of a servo system. The servo system is an electromechanical system that positions a mechanical device using electrical motors. The characteristics of a servo system will be reviewed later in this chapter.

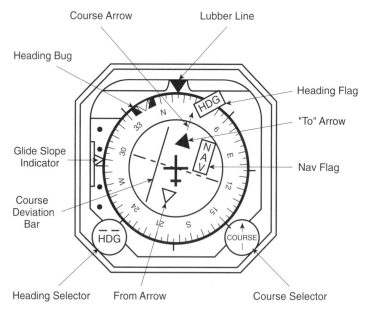

Figure 8–7   Example of an HSI.

To better understand the HSI, let us study an example. Assume that the OBS is set to 45° TO, which means that the arrow is pointing toward the 45° mark on the compass card, the CDI is centered, and the aircraft heading is 0°. Therefore, the heading card is oriented such that 0° is at the top of the card with the 0° marking just below the pointer, called a *lubber line,* at the top of the instrument. The OBS arrow is at an angle of 45° from the vertical, with the head of the arrow pointing to 45° on the compass card.

The horizontal situation is clear. The aircraft is on a heading of 0° the VOR is to the right, and the aircraft is on the selected radial, 45°. The aircraft icon on the HSI shows that the aircraft is not flying the radial, and if the 0° heading is continued for any length of time, the selected radial will be behind the aircraft. This will be indicated by the movement of the center section of the OBS arrow. If the aircraft is turned to the right so that the heading and the selected radial are the same, the heading will be 45°, the aircraft will be heading directly to the VOR indicated by the head of the arrow, and the CDI will remain centered. The HSI will show the course arrow pointing straight up at the 45° mark on the compass card below the lubber line and the CDI line segment in the center of the arrow, where it will remain.

If the aircraft is turned in a tight turn so that it does not stray from the selected radial to a heading of 225°, the entire compass card, CDI, and OBS will all rotate 180°. The HSI now displays an aircraft, still on course, with a heading of 225°; the selected radial is still 45°, but with the arrow pointing to aft of the aircraft in the direction of the VOR. The horizontal situation is that the aircraft is flying on the selected radial, but flying away from the VOR.

Compare this maneuver with a conventional OBS–CDI indicator. As the aircraft is turned, the directional gyro will change from 45° to 225°, just as with the HSI, but the CDI and TO/FROM indicator will not change, as explained in Chapter 5. Now the aircraft is still flying the 45° TO radial but is headed away from the VOR. Therefore, the sense of the CDI is reversed and does not provide the correct information; it is the reverse of what it should be. The correct procedure requires that the OBS be changed to 225°, when the TO/FROM flag will change to FROM and the steering information will be correct. In the HSI, however, the CDI rotated 180° with the compass card, which automatically reversed the sense of the CDI, which continues to provide the correct steering information.

The HSI makes it easier to intercept a selected radial. The movable section of the radial arrow represents the actual distance from the selected radial. If the CDI is deflected such that the line is ahead of the aircraft icon, the radial will be intercepted. If the CDI is deflected such that the movable section is behind the aircraft icon, the radial will not be intercepted unless the aircraft is turned.

There is a subtle, but important difference between the CDI for an HSI and the CDI used for a conventional VOR indicator. The conventional CDI shows the angular error of track. As an example, a full-scale deflection is 10°, which is a much greater distance from the selected radial at 100 nautical miles (nmi) from the VOR than it is at 10 nmi. To provide a pictorial display, the CDI must provide a deflection proportional to the distance to the selected radial. This requires that the HSI signal converter have DME information. Most HSIs provide a display of DME distance in the upper-left corner of the instrument.

To summarize the HSI, the aircraft icon is your aircraft; you are heading in the direction indicated by the number below the lubber line, the VOR station is in the general

Photo 8–2   A Horizontal Situation Indicator which provides the aircraft heading, bearing, course, course deviation, VOR TO/FROM, and glidescope division. The unit may be switched to perform as an RMI. Photo courtesy of Aeronetics, a division of AAR Manufacturing, Inc.

direction pointed to by the course select arrow, and the selected radial is the movable section of the arrow.

Example

How is the DME distance used to provide the CDI deflection for an HSI?

Solution.   The distance to the selected radial is one side of a right triangle consisting of the DME distance, which is the hypotenuse, the distance from the VORTAC to the desired position on the selected radial, and the distance to the desired course as shown next.

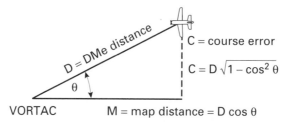

First we write the equation for the triangle. The distance, *D,* is a known because this is the DME distance. The angle θ is also a known because it is available from the VOR receiver as the conventional CDI deflection. The equation is solved for the short side of the triangle or the distance from the desired VOR radial.

If an area navigation computer is in use, the information will be relative to a waypoint rather than a VOR. The HSI is also used for any of the long-distance navigation systems that calculate latitude and longitude by calculating the course error for a selected radial and a waypoint.

The HSI may be used for landing by providing localizer information. To land using the HSI, the heading of the runway is set using the OBS knob. The OBS is not active in the localizer mode, and the OBS arrow is used more as a cursor. There is no TO/FROM indi-

Photo 8-3   An HSI with a remote mounted gyro. Photo courtesy of Aeronetics, a division of AAR Manufacturing, Inc.

cation either, and the head of the OBS arrow is permanently in the TO direction. Normal landing procedures calls for the aircraft to be turned to the selected heading, and the center segment of the OBS arrow will then provide the left–right deviation information from the localizer receiver. The arrow is pointing straight up, the aircraft heading is the same as the runway's, and the segmented section of the course arrow is centered.

If the aircraft departs using the localizer as departure guidance, the course deviation indicator continues to provide valid guidance from the back course. If the aircraft makes a 180° turn and approaches the departure end of the runway using the back course, because the compass card and the CDI that rotates with the compass card have both rotated 180°, the CDI deviation has reversed, which automatically compensates for the reversed course deviation sense on back-course approaches. The runway heading remains set by the OBS because the normal approach heading and the arrow now point down, which indicates that the aircraft is approaching the departure end of the runway.

A glide-slope indicator is provided as part of the HSI so that complete ILS information is available from the HSI for instrument landings. The glide-slope indicator is a vertical pointer to the left of the HSI display.

SERVO SYSTEMS

Servo systems are used extensively in indicators and in the flight control systems discussed in Chapter 9. It is necessary to have an understanding of servo systems to fully appreciate these instruments.

There is a huge variation in servo systems, and those most often encountered in indicators and flight control systems provide angular motion. A *servo system* may be mechanical, electrical, hydraulic, or combinations of these and controls a parameter relative to an input. A servo system is a feedback system that compares an input with the output and supplies gain to control the output device until it is the same as the input.

In electromechanical servo systems the position of some mechanical device is controlled electronically. Often the input to a servo system is a mechanical signal that is processed electronically and results in a mechanical motion.

Typically, the servo system adds torque to the input signal in order to position a larger, more massive, mechanical assembly. An example of a servo system is the common power steering system for an automobile. The input to the system is the steering wheel and the output is the front wheels. The input requires a small torque, whereas the motion of the wheels may require considerably greater amounts of torque. Although this servo system is all mechanical, using hydraulic power, it serves as a well-known example.

The block diagram of an electrical angular servo system is shown in Fig. 8-8. The input, a mechanical shaft rotation, is applied to a servo motor. As previously explained in Chapter 5, a servo motor does not always provide mechanical shaft rotation from an electrical input as a conventional motor. The desired output shaft is coupled to a similar servo motor. The outputs of the two servo motors are compared by subtracting one from the other, producing an *error signal.*

A very common servo motor used in aircraft servo systems is the *synchro,* which is an important part of angular servo systems. This device is similar to the resolver described in Chapter 5, which was used to provide phase shift as a function of shaft angle for the VOR–OBS.

The difference between the synchro and the resolver is that there are three stator windings in the synchro compared to the resolver's two. Usually, the rotor is powered from a 400-hertz (Hz) ac source, and the outputs of the three stators are functions of the shaft angle. The three stators are separated by 120°, and the output voltage is proportional to the sine

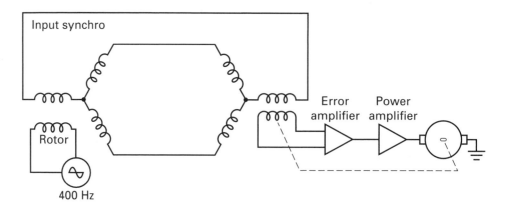

Figure 8–8   Torque-amplifying servo system.

of the angle between the rotor and the stator. Like the resolver, the synchro is a precision motor, has very fine quality bearings, and is made with great precision.

In this example, the output shaft is connected to an indicator such as an HSI or a directional gyro indicator. The indicator can be located away from the gyro and the gyro and indicator connected with electrical connections. Also, other synchros may be connected to the gyro indicator shaft and used as input signals for other servo systems to drive other indicators. As an example, the servo system could turn the directional gyro indicator and two additional synchros to drive an HSI and an autopilot. These additional synchros are called *repeater synchros.*

When both synchros have the same shaft angle, the difference in their output voltages is zero, and when the shaft angles are different, the difference is called an *error voltage.* This error voltage is amplified and used to drive the synchros to the same shaft angle using another motor, not necessarily a precision motor.

The subtraction process can be implemented by using a differencing amplifier or, more simply, by connecting the input and output servo motor stators together with the correct phase relationship so that the rotor output voltage is at a null when the shaft angles are 90° different. This 90° offset is easily compensated for by offsetting one of the synchros by 90°.

The rotor of the input synchro is powered from a 400-Hz ac source, which makes the input synchro the transmitting synchro. The three stators of the input synchro are connected to the three stators of the feedback synchro. The feedback synchro is mechanically coupled to the output shaft. If the input and feedback synchros are set to exactly the same shaft positions, the voltage from the rotor of the feedback synchro is the same as the input's rotor. If the feedback synchro is rotated 180°, the amplitude of the signal will be the same, but the polarity will be reversed, which is the same as 180° out of phase. If the shaft angle is 90° different, the output of the receiver synchro will be zero. Notice that two orientations of the receiver synchro are 90° different from the transmitter synchro. Figure 8-9 shows the output synchro's rotor voltage as a function of shaft angle.

In the example of Fig. 8-8, the error voltage from the feedback servo motor is amplified and used to provide the drive for the motor, which causes the output device to rotate until the feedback servo motor has a zero error voltage. Any change in the input or feedback servo motors causes the error voltage to increase and the driving motor to turn to correct the error. The motor will continue to turn until the feedback servo motor is driven to a null, or its point of zero output.

Two shaft positions provide a null voltage from the feedback synchro, but only one position is correct. One null position provides an output that has a polarity for off-null conditions that tends to cause the servo system to return to the null. This is a condition of negative feedback, or a polarity of feedback that keeps the servo system nulled. The other null position is such that the motion of the servo system is away from the null and toward the correct null 180° away. Even the slightest disturbance causes the servo to move off the undesired null and toward the correct null.

When the motor has driven the output shaft to the correct null, the feedback synchro shaft is offset from the input shaft by 90°. Correcting this situation is a simple matter of aligning the input and feedback synchros with a 90° offset.

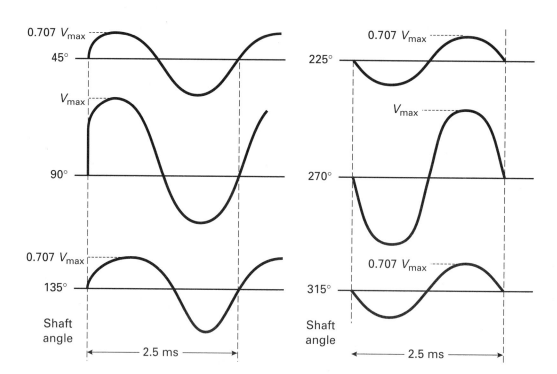

Figure 8–9   Relationship between shaft positions and the feedback synchro rotor voltage.

The servo system can provide angular movement or linear movement. Most servo systems found in aircraft provide angular movement.

One important application of a servo system is torque amplification. An example of this is coupling a directional gyro to the rotating compass card of an RMI or HSI. The frictional forces of these indicators causes excess gyro drift. By coupling a very light load, such as a synchro, to the gyro, the gyro drift can be kept to a minimum. The synchro becomes the input to a servo system that positions the compass card of the HSI or RMI to the same rotational orientation as the gyro. The output of a servo system can provide as much torque as needed for such applications as turning something as massive as a gun turret or as delicate as an airborne indicator. The signal from the gyro synchro can be distributed among the instruments aboard the aircraft that can use heading information, such as HSIs, RMIs, and autopilots.

A similar synchro-based servo system is applied to the vertical gyro to provide roll and pitch information. To supply both requires that two synchros be employed. These signals are used by the autopilot, for driving instruments such as the attitude display indicator, and for stabilizing the weather radar antenna.

INDICATOR ELECTRONICS

Indicators such as the HSI rely on a large variety of servo systems for operation. Another important component of the more sophisticated instruments is the slaved gyro.

In the *slaved gyro* the necessary corrections to compensate for drift are done automatically. An electronic magnetic compass is required to provide the necessary information to correct the gyro for drift. This device is called the *flux gate compass.*

A flux gate compass uses the magnetic field switching characteristics of a ferromagnetic material. Figure 8-10 shows the shape of the flux gate. The frame is made of a ferromagnetic material that concentrates magnetic flux, as explained in Chapter 5 concerning ADF antennas. Ferromagnetic material has another characteristic, *saturation,* that will allow the magnetic flux to be switched. When a certain magnetic field intensity is reached, the ferromagnetic material can support no more magnetic flux. When a ferromagnetic material is saturated, magnetic lines of flux surrounding the material are no longer attracted to it, and it has essentially lost its ferromagnetic characteristics.

The structure shown in Fig. 8-10 contains an excitation coil that magnetizes the entire structure to saturation. The excitation is 400 Hz, which implies that saturation exists twice for each 400-Hz cycle, once for each positive half-cycle and again for the negative half-cycle. Thus, the entire structure goes from ferromagnetic to nonmagnetic 800 times a second.

When the flux gate is not in saturation, the lines of flux pass through the structure as shown in Fig. 8-11. Notice that the amount of flux passing through any one of the three legs is a function of the angle between Earth's magnetic field and the flux gate.

Three pickup coils are mounted on the three legs as shown. Because the magnetic field is switched on and off by the entire structure being driven from ferromagnetic to nonmagnetic, there are 800 flux changes in each pickup coil, where the amount of flux is a function of the angle between Earth's magnetic field and the flux gate structure.

The flux gate is a device that permits the magnetic flux to pass through the legs of the structure or to bypass the structure, much like opening and closing a gate. Because the structure has three legs with a 120° angle between each leg, the amount of magnetic field orientation is the same as would be found in a synchro with the rotor generating the magnetic field and the stators as pickup coils. This produces an output voltage at each pickup coil at a frequency of 800 Hz for which the amplitude follows the same angular relationships as found in a synchro.

Because of this similarity to a synchro, the output from each pickup coil is the same as a synchro, and thus the flux gate pickup coils can be used as the equivalent of an input synchro in a torque-amplifying servo system. The flux gate is connected to a feedback synchro that is connected to the directional gyro with the necessary 90° offset. The null in the feedback synchro occurs when the directional gyro has the same angular orientation as Earth's magnetic field relative to the flux gate. In this example the frequency of the flux gate synchrolike signals is 800 Hz rather than the usual 400 Hz, and no shaft is associated with the flux gate.

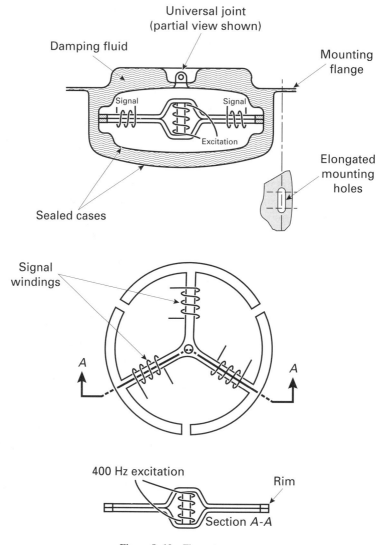

Figure 8–10 Flux gate compass.

To complete the servo system, a motor must be provided that will move the gyro to create a null from the feedback servo. Remember that a gyro's main purpose is to resist movement. However, the amount that the gyro must be moved to correct for drift is very small, and a system called a *gyro torquer* provides the motion. The torquer is a simple motor mounted to the gyro gimbal that provides a steady source of torque, which produces a slow but adequate motion of the gyro.

Usually, a flag is provided that detects when the gyro and the flux gate differ by large amounts, which occurs when a large amount of torque is applied to the gyro. This flag is

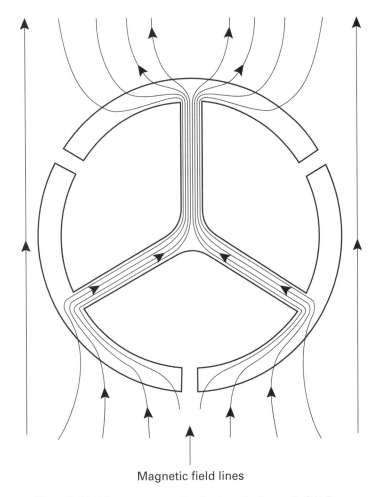

Magnetic field lines

Figure 8–11   Flux gate compass showing the path of magnetic field lines.

called the *heading flag* and serves the same purpose as a nav flag by indicating that the heading information may not be reliable.

The flux gate should be mounted away from large magnetic fields such as those found near an engine or electric motors. Typically, a flux gate is mounted near the end of a wing or far aft on the fuselage away from electrical apparatus.

A typical directional gyro installation uses a remote mounted gyro that provides heading information for several aircraft systems. The gyro is typically mounted near the center of gravity of the aircraft to minimize the amount of displacement due to bank and pitch changes. The directional gyro can have one or more synchros to provide heading information for the HSIs, RMIs, or autopilots.

INERTIAL NAVIGATION SYSTEM

Although gyros have been used for many years and the quality and cost of aircraft gyro systems have steadily improved, the gyro is a heavy, mechanical device. A method of navigation that does not require a radio signal is called an *inertial navigation system,* or INS. The INS uses transducers to measure the acceleration of the aircraft in all three axis.

When acceleration is mathematically integrated over time, the result is velocity. When velocity is integrated, again over time, the result is distance. Accelerations are measured using a mechanical device called an *accelerometer.* Accelerations are measured on all three orthogonal axes, which are integrated to find velocity and distance, also on the same three axis. For the INS to work, the orientation of the accelerometers must be very precise. The accelerometers are mounted on a device called an *inertial platform* that provides an orientation for the *X, Y,* and *Z* axes regardless of the aircraft's maneuvers. However, it is important that the platform maintain an accurate orientation to the *X, Y,* and *Z* axes, a process called *stabilization.* Platform stabilization is achieved by using gyros as position references.

To stabilize an inertial platform, three gyros are used, one for each axis. The accuracy of the inertial navigation system depends on the gyros remaining oriented and not drifting. An error in the orientation of one axis of an INS will cause an error in navigation accuracy. The precision of the stabilized platform is very important, because the position of the aircraft is determined from integrating an acceleration in one of the three axes. The error is cumulative; if an acceleration is incorrect, this flawed value is used for a long period of time to derive the aircraft's position, which results in a large and continually increasing error. The stabilized platform is used to mount the accelerometers for the inertial reference system. The stabilized platform can also provide roll, pitch, and yaw information for the instruments requiring this information.

The INS requires a stable platform that involves three gyros and three accelerometers. Gyros tend to have drift and accelerometers have offsets. Because integration over time is involved for determining position with an INS, these error sources produce a cumulative effect. Like the directional gyro, the INS must be corrected for drift. Another problem is that the INS, theoretically, provides a platform that is independent of the shape of Earth. In other words, if an aircraft is flown in a straight line, relative to a perfect INS, it will fly tangentially into space.

Two of the three axes of the stabilized platform, the roll and pitch, can be corrected by sensing the direction of the gravity vector. When the aircraft is not under accelerated motion, which is during straight and level flight, which represents the majority of the aircraft's flight time, the roll and pitch axes can be corrected for drift. Correcting drift in the yaw axis requires an outside reference, because the aircraft does not fly a specific yaw axis. To correct for yaw axis drift, information from a radio navigation system such as omega or GPS is used.

The INS without correction is a valuable navigation aid, but without long-term correction for drift, it becomes useless after a period of time. The advantage of the INS is that it provides instantaneous position information in three dimensions. Long-term information is actually from other navigation aids such as a GPS or omega.

Inertial navigation systems are usually found only on large transport aircraft where the directional gyro and the artificial horizon are replaced with an inertial reference unit, or IRU. When one element such as the IRU provides so much of the aircraft with vital information, the unit is backed up by a second system. In addition, the IRU has a battery backup so that a loss of primary power will not cause the IRU to tumble and lose its orientation.

LASER GYRO

The stability of the INS platform is in direct relation to the quality of the gyros used to stabilize the inertial platform. An improved quality gyro has been developed that greatly enhances the performance of the INS. This is the laser gyro, which uses the effects of counterrotating laser beams. This gyro, which is sometimes called the laser ring gyro, is shown in Fig. 8-12. A triangular tube is filled with a helium–neon gas mixture, which has the characteristics needed to create laser light. All three corners of the triangle have mirrors, which cause the light energy to traverse the triangle in either direction. The gas in the tube is subjected to an electric field that ionizes the gas and places the gas in a higher-energy state. One possible elevated energy state is a metastable state. This is an energy level at which the gas can remain for some time. When the gas returns to the lower-energy state, a photon of light is emitted with a wavelength for that energy state. In the case of a metastable state, the gas molecules can be stimulated to emit a photon of light by another photon. Therefore, a coherent wavefront of photons is generated that produces a very pure light of only one wavelength. The name *laser* is an acronym for Light Amplification through Stimulated Emission of Radiation. Essentially, the device is a Ne–He laser with the ends bent so that they meet at one corner of the triangle. Lasers require that the light energy reflect between the ends of the laser to achieve laser action. In this laser, the ends are brought together.

Where the ends of the laser are brought together, a corner reflector is provided so that the laser beam from the two ends of the bent laser falls on the same photo detector. If the light is in phase, the two light beams will combine to produce a bright spot. If the light beams are out of phase, they will combine to produce a dark spot.

When the laser gyro is rotating, the distance traveled by the light energy in the clockwise direction is different from the distance traveled by the light energy in the counterclockwise direction. Thus the number of light waves in the space around the laser gyro for the clockwise path is different from the counterclockwise path. Therefore, the light combining at the photo detector will change from dark to bright depending on the number of wavelengths of difference between the clockwise and counterclockwise paths.

When the laser gyro is used to stabilize a platform, the change of brightness at the photodetector is proportional to the rate at which the laser gyro is rotating. These changes are detected and used to drive a motor that resists the rotation of the platform and restores the platform to its original position.

The laser ring gyro senses rotation about an axis passing through the center of the gyro. Therefore, to stabilize a platform, a ring laser gyro is attached to the shaft of a gimbal to sense rotation of that gimbal. Torquers are used to return the platform to its original position when forces are applied to the platform. Most of these forces are due to the friction of the platform bearings and any acceleration-induced forces.

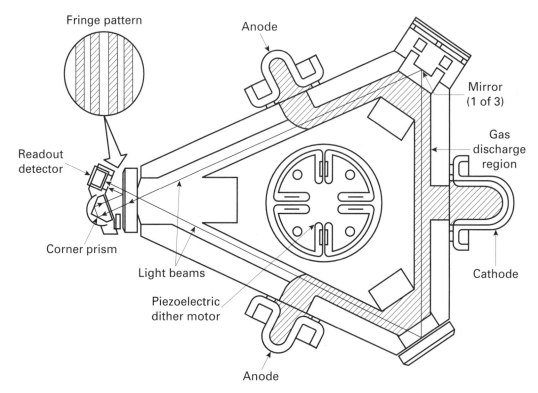

Figure 8–12   View of the laser gyro.

The laser gyro has a number of advantages over the mechanical gyro. First, it has no moving parts and thus should be more reliable. Also, not requiring a large, massive rotating wheel, the laser gyro is lighter. The laser gyro is very sensitive to rotational motion and can provide an excellent stable platform for an INS.

ELECTRONIC FLIGHT INSTRUMENT SYSTEMS

The mechanical display instruments that have been presenting information to pilots and crew for as many as 50 years have some significant disadvantages. First, because they are mechanical instruments, they are less reliable than other display technologies. They have become increasingly more expensive because considerable skilled manual labor is required to assemble these instruments. Third, they are large because the more sophisticated instruments contain an number of synchros, motors, and other parts. Finally, the instruments lack flexibility because one instrument displays only one feature.

It is desirable to include more features in one instrument to present a more comprehensive display. As an example, the HSI is a combination radio navigation and gyro-based instrument that creates a pictorial display. But this instrument is a very complex and expensive unit. For a number of years there has been a push to develop electronic display technologies to provide flight instruments.

Some mechanical displays were easily updated with electronic displays. The rotating drums that displayed frequency and DME distance were easily replaced with seven-segment digital readouts. The more complex instruments are more difficult to replace.

DISPLAY TECHNOLOGIES

Two basic types of displays are used in aircraft, alphanumeric and graphics. Alphanumeric displays display simple text-type information, such as the frequency of a radio or the squawk for a transponder. Graphics displays are pictorial types of displays, such as an electronic HSI or ADF indicator.

The most common alphanumeric display type is the seven-segment display for numeric data. The seven segments do not produce the most artistic of numeric shapes, but the numbers are easily seen and provide an unambiguous display. The seven-segment display cannot produce letters, and modifications to the seven-segment display are made to produce alphabetical characters. These characters are less than ideal, and most alphanumeric data are provided by either a 5 × 7 or a 7 × 9 matrix of dots. Examples of some numeric characters are shown in Fig. 8-13.

There are two basic techniques for creating electronic data displays: light emission or reflective. Both techniques have been used for a very long time and do not represent new technology unique to electronic displays. An example of a common emissive display is a *lighted annunciator.* This is a message that is printed on a translucent panel with a lamp mounted behind the message. When the lamp is illuminated, the message is easily read. An example of a reflective display is the nav flag on an OBS–CDI indicator. When the flag message is to appear, a printed OFF appears in a window.

Both types of display techniques have their advantages and disadvantages. The reflective technique must have a source of illumination to be viewed at night. Of course, many mechanical instruments, such as the OBS–CDI, must be illuminated at night, and a system of instrument panel lights is very common in aircraft.

The emissive display has the opposite problem. Since it is an illuminated display, it does not need to be illuminated at night. But the amount of light intensity to be seen in the very high ambient light of direct sunlight is very great. When sufficient light intensity is provided for daylight viewing, the level of intensity is excessive for nighttime viewing and the intensity must be adjusted.

Figure 8–13   Examples of numeric displays.

A graphics display is an extension of the matrix display used for alphanumeric data displays. A graphics display is divided into basic picture elements called *pixels.* Each pixel has associated with it a number of characteristics that are essential for generating the picture. If the graphics display is a simple line drawing, the pixel will have two values, on or off, light or dark, or from a logic standpoint a 0 or 1. If the display has a gray scale, that is, different shades of gray, the pixel will have a number of values depending on the possible shades of gray. If the display is a color display with a gray scale, the pixel will have the intensity for each primary colors, red, green, and blue. Essentially, any graphics display is generated by an array of dots that produce the picture. Unless the display is viewed with a magnifying glass, the pixels cannot be seen. To give some idea of how many pixels are required, a 5-in. display with an array of 256 by 256 pixels has visible dots. This display is not objectionable but the dots are evident. A 512 by 512 array has pixels that are barely visible, and in a 1024 by 1024 array the pixels are effectively invisible. Of course, a larger display requires larger pixels, and therefore a 5-in.^{2} display with 512 by 512 pixels may be satisfactory, but in a 10-in. display with the same number of pixels, the larger pixels will become visible.

A number of display technologies have been and continue to be used in aircraft systems. One of the first display technologies was *light-emitting diodes,* or LEDs. Some characteristics of LEDs were discussed in Chapter 2. The first available LEDs were only red in color, and were not readily accepted in aircraft because red is reserved for warnings. When green, orange, and yellow LEDS became available, they were mostly used for numeric-type data, such as frequencies, clocks, and DME distances.

LEDs emit light, and thus the light intensity must be greater than full sunlight in order for the display to be seen in full sun. Quite a bit of power must be applied to an LED, in addition to providing color filters and some hooding of the display to prevent direct sunlight striking the display, to make it sunlight readable. LEDs are used for some graphic data, but large LED displays tend to consume considerable power, are expensive, and, until recently, could not provide full color because of the lack of a blue LED.

Liquid crystal displays (LCDs) are not emitters of light but serve as light shutters. An LCD display consists of a liquid crystal material sandwiched between two plates of glass. One type of liquid crystal display has metallic material applied to both pieces of glass, with the desired display pattern on the front piece of glass and a solid metal surface on the rear. The metal is very thin so that it transmits light with very little attenuation. When an ac electric field is applied between the front and rear plates, the liquid crystal material becomes opaque to light.

There are several types of liquid crystal materials. Some require the use of polarized light, which is generated using polarizing filters; other materials work with random polarization. All liquid crystal material acts as a light shutter.

The liquid crystal display can operate in two modes. First, a light source can be placed behind the display and the LCD acts as a light shutter. This is called a *transmissive display,* meaning that it is an emissive display but the transmission of light is controlled by the LCD shutter. The problem associated with this type of display is that the light source must be sufficiently bright to ensure that the display is sunlight readable.

Another mode of operation of the LCD is to provide a light-colored background behind the LCD so that light is reflected from the background if the LCD light shutter is open. This type of display is a *reflective display* and, like all other reflective displays, has no sunlight readability problems because the sunlight is reflected to create the display.

The problem with a reflective display is that it cannot be seen at night. A simple solution to this problem is to provide a translucent background to the LCD and a light source for use at night. This mode of operation is called *transflective* and is used in most aircraft LCD displays.

The LCD has a number of advantages. First, the LCD light shutter is a very low power device. No power is required to darken an LCD shutter. If the display is used in the reflective mode, the power consumption is very low. When a light source is added for night viewing, the intensity of the light source is low because there is no bright sun to overcome, and so power consumption is still low, even with the added power required for backlighting.

LCDs can provide graphic data displays. This is accomplished by using an array of a large number of LCD light shutters. When monochrome graphics are required, the LCD may be used in either the reflective, transmissive, or transflective mode.

Color graphics can be displayed with LCD technology by providing three light shutters for each pixel. Each shutter is backed up by a red, green, or blue filter so that each shutter provides one of the three primary colors.

The color display must operate in the transmissive mode because the light must pass through the filter to generate the three primary colors. To be sunlight readable, the intensity of the backlighting must be able to overcome direct sunlight. This requires an intense light, but this is a simple device with reasonable efficiency.

One problem with the LCD display, particularly the color display, is that an alternating electric field must be selectively applied to each pixel. With the hundreds of thousands of pixels involved even with a small display, it is virtually impossible to provide a connection for each pixel. One common method of addressing a large array of elements is to address the pixels by row and column, and where the rows and columns intersect, an electric field may be generated. The main disadvantage of this technique is that the pixel is addressed for only a very short time, and the applied electric field will be present for a very short time. This results in a serious loss of display contrast, which is not acceptable. A technique used to overcome this problem is to provide a transistor switch for each pixel and address the transistor switch, which will apply the electric field to the pixel until the switch is addressed again. A typical transistor switch is a MOSFET; the gate capacitance is charged to a point where the MOSFET conducts and provides the necessary electric field across the pixel. The charge applied to the gate only needs to remain until the transistor is addressed again. The problem is that the transistors must be sufficiently small to ensure that the transistors do not block the light from the pixel. Transistor technology can provide small-geometry transistors that will fill this need. The devices are called *thin-film transistors,* or TFTs.

TFT LCD displays provide remarkable performance and show much promise for future displays. The problems at the time of this writing are manufacturing yields, because a display cannot tolerate any defective transistors. It is very difficult to fabricate nearly 100,000 transistors plus the necessary interconnecting wiring without a failure. Thus, manufacturing yields are low and the result is a high price for color LCD displays.

A significant advantage of LCD graphics display is that it can be very thin and require very little space behind the panel, thus producing what is called a flat panel display.

Even though the color LCD display contains the three primary colors, LCD shutters have an on and an off state, which implies that the colors available from an LCD display are those colors that involve equal amounts of primary colors. The colors available from an LCD are the three primaries, red, blue, and green; white, which is all three primaries combined; blue and green in equal amounts, which is cyan; red and blue in equal amounts, which is magenta; and red and green in equal amounts, which is yellow. The use and generation of these colors is discussed further in connection with cathode-ray tubes.

There are other problems associated with LCDs, one of which is temperature. The LCD cannot operate at temperatures below about $-20°C$. The LCD can be permanently damaged at temperatures well below this. The LCD action slows to a point where it is virtually unworkable at low temperatures. However, this is not a typical ambient temperature for very long in most cockpits.

Another problem with LCDs is that they are made of thin glass and tend to be fragile. This problem is common for many electronic components, and the LCD must be mounted and protected from undue stresses and impacts.

The major display technology used for Electronic Flight Instrumentation System (EFIS) systems at the time of the writing is the venerable *cathode-ray tube,* or CRT. Perhaps the first introduction of the CRT into civilian aircraft was in the late 1950s, when the first airborne weather radars were installed in aircraft. There have been a vast number of improvements to the CRT, but many of the major disadvantages persist. Probably the biggest problem with any CRT is its depth behind the panel. With the wide-deflection CRTs in use today, the required depth is nothing compared to what it once was, but there is still considerable depth.

Another problem with the CRT is the large number of different operating voltages required, particularly the high voltage. These voltages require extensive power supplies, and the presence of high voltage is always undesirable in airborne equipment because of the problems with arcing at a high altitudes.

One significant advantage of the CRT is that it is a very mature technology. Because of the huge number of CRTS manufactured for television receivers, manufacturing technology is very advanced and CRTs are inexpensive. Also, effective circuits are available for powering and controlling CRTs.

Figure 8-14 shows the basic construction of the CRT. The CRT uses an evacuated glass tube. At the rear of the tube is an electron gun that accelerates and focuses a beam of electrons onto a phosphor screen. The electrons are made available for acceleration to the front of the CRT by heating a cathode of a metal, where added energy from the hot cathode allows the electrons to be separated from their atoms. An electric field is applied to the cathode so that the electrons set free by the heat will move in the direction of the electric field and away from the cathode. The electrons are accelerated through several fields so that they reach a very high velocity and head directly toward the CRT screen. When the electron beam collides with the phosphor atoms, the phosphor emits light.

The color CRT has three electron guns that strike a screen made of narrow stripes of different phosphors that emit three different colors. The three colors are the three additive primaries, red, green, and blue. The electron beam is assured of striking only one phosphor color by directing the electron beam through a slit in a shadow mask so that the desired color phosphor is struck.

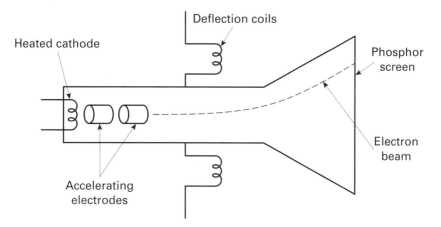

Figure 8–14   Construction of a CRT.

The amount of current in each electron gun beam determines the intensity of light emitted by the color controlled by each beam. Therefore, the intensity of each of the three primaries can be controlled and thus any color possible may be generated.

The electron beam is directed along the center of the CRT, and unless the beam is deflected by the application of transverse electric or magnetic fields it will strike only the center of the CRT screen. The electron beam in an aircraft display CRT is deflected across the phosphor screen by a magnetic field because the magnetic field can deflect the beam through much greater angles than an electric field, and thus a shorter CRT may be employed. As the beam is deflected, the electron beam may be turned off and on to write on the phosphor screen.

There are two methods of generating the image on the screen. The first is called *stroke scan* and operates just as an artist would in making a picture. As an example, if the graphics display is to be of an ADF indicator, the numbers are drawn on the screen one at a time, just as an artist would with a pencil. After the numbers, the graduations are placed around the outside of the display.

This technique has difficulties when large solid areas must be drawn, such as the blue sky segment of an attitude direction indicator. To make this type of display with a pencil, the pencil is repeatedly drawn back and forth to fill in the area. This type of back and forth motion is the basis behind the raster scan technique of display generation.

The *raster-scan* technique is the method used to generate television pictures. The entire picture to be generated is scanned from the top left side to the bottom right side by lines. The picture is generated a line at a time.

Just as there must be enough pixels so that the pixels are not easily seen in the display, there must be enough scanning lines so that the display does not show the lines. A U.S. television picture has 525 lines, and many high-resolution displays have 1024 lines.

It is easy to calculate how many pixels are required for a scanned display. The technique used is to imagine the display constructed of square pixels. If the display is generated using 1024 lines, there would be 1024 rows of pixels. To determine how many pixels on each scanning line, the aspect ratio of the display must be known.

The *aspect ratio* is the width divided by the height, assuming that the scanning lines are horizontal. (The scanning lines do not have to be horizontal but usually are.) As an example of a familiar aspect ratio, U.S. television pictures have a 4:3 aspect ratio, and most motion pictures have a 16:9 aspect ratio. A number of aspect ratios are used in aircraft display systems, but 1:1 is common.

Pixels are considered to be square, having equal horizontal and vertical dimensions. Therefore, the number of pixels on a scanning line is the aspect ratio times the number of scanning lines. The total pixels for an entire display are the sum of all the pixels on the scanning lines or

$$(\text{number of lines})^2 \times \text{aspect ratio}$$

Example

How many pixels are there in an airborne display with 1024 lines and a 4:3 aspect ratio?

Solution.   The number of pixels is

$$\frac{4}{3} \times 1024^2 = 1.4 \times 10^6$$

Obviously, there are a lot of pixels in a display.

The number of storage bits required to store an entire display picture depends on the number of attributes associated with a pixel. The simplest type of display, such as the HSI, requires only one attribute per pixel, on or off, because the HSI is created with monochrome lines. If the display is a color display, each pixel will have to control three primary colors, which will require at least three bits, one for each color. If a gray scale is present, that is, there are different levels of each color, rather than just on or off, considerably more bits are required. If, as an example, a display had 16 levels of gray scale, 4 bits are required for each color or a total of 12 bits per pixel.

Sixteen levels of color are the requirements for displays such as color television, and most aircraft applications require only an on/off function for each color. A total of seven different "colors," which includes white as one of the colors, may be generated with 1 bit per primary color. This is sufficient for airborne color displays. The colors that may be generated are shown in Table 8-1. A zero indicates that that color is not illuminated, whereas a one indicates full brightness for that color. Returning to our example of a 4:3 display with 1024 lines, the total memory capacity to store the display is 1.4×10^6 pixels times 3 bits, 1 bit per color, or 4.2 million bits.

The CRT display must be scanned rapidly enough that the display appears to be continuous. The phosphor of the CRT has some decay time, but this is slower than the retentivity of the human eye. The decay time of the human eye is such that, if the display is redrawn about every 1/50th of a second, it will appear continuous. If the display is updated at a slow rate, the display will appear to flicker. As a benchmark, a television picture is updated one-half a picture at a time by scanning first the odd lines and then the even lines every 1/60 second(s). This results in a complete picture updated every 1/30 s. In the case of an aircraft CRT display, there is no rapid motion and updates can occur less often than for a television display. Typically, 50 times a second is the minimum for updating the CRT display.

TABLE 8-1

Color	R	G	B
Black (dark)	0	0	0
Red	1	0	0
Green	0	1	0
Blue	0	0	1
Yellow	1	1	0
Magenta	1	0	1
Cyan	0	1	1
White	1	1	1

Example

At what rate must the video RAM be accessed to provide the storage of a CRT display? The picture to be displayed requires the colors shown in Table 8-1 and has a 4 : 3 aspect ratio and 1024 horizontal lines.

Solution. If an entire picture is redrawn 50 times a second, the entire memory must be accessed in 1/50 s, which implies that 4.2 million bits must be accessed or at a rate of

$$\text{rate} = 4.2 \times 10^6 \times 50 = 210 \times 10^6 \text{ bits per second}$$

This implies that a bit must be available from the memory at a rate of one every 4.8 nanoseconds, (ns), of which no memory is capable.

It would be prudent to arrange the memory such that the 3 bits required for each pixel are available together. This is called *memory arrangement,* and the memory for our example CRT display could be arranged as 1.4 million words by 3 bits. Since we are now accessing the memory by words, one word per pixel, the access rate is $1.4 \times 10^6 \times 50 = 70 \times 10^6$ words per second. This requires that a word be available every 14 ns, which is possible but still a fast memory.

Why not arrange the memory as a 9-bit word and extract the information for three pixels from the memory at a time? This is commonly done, and the access time for the memory is now 52 ns, which is a very reasonable memory speed.

The raster scan display requires that, regardless of the nature of the CRT display, the entire CRT screen be scanned. This requires the electron beam to travel every line of the display once for each generated picture, a long distance for each picture. Thus, the beam current must be increased to provide a bright display. The brightness of the display is proportional to the current but inversely proportional to the beam deflection rate.

Example

What is the beam velocity of a 10-centimeter-square (cm^2) CRT having 1024 lines and an update rate of 1/50 s?

Solution. Ignoring the retrace lines, 1024 lines of length 10 cm are scanned 50 times per second; therefore, the beam velocity is

$$V = 1024 \times 0.1 \times 50 = 5120 \text{ meters per second (m/s)}$$

Figure 8-15 shows a block diagram of a CRT display generator. The microprocessor receives the input data to be processed. Using the HSI example, the input information is the heading of the aircraft, the OBS setting, and the CDI information. The heading is used to set the orientation of the compass card. As previously explained, the compass card appears the same regardless of the heading of the aircraft; it is only the orientation of the compass card that is a function of the aircraft heading.

The microprocessor retrieves the image of the compass card from the ROM, calculates the coordinates of the pixels for the orientation required by the current heading, and stores the new image in the video RAM. Because the compass card involves angular motion, the best format to store the image is as polar coordinates. Remember, also, that the compass card requires that only those pixels that are illuminated need to be installed; the remaining pixels are off. With a display such as an HSI, less than 5% of the display is illuminated, and of the more than 1 million pixels in a 1024 by 1024 square display, only 50,000 pixels are involved. If the illuminated compass card pixels are stored as polar coordinates, creating a rotated version of the image is a simple matter of adding the heading angle to the angle of the coordinate of all stored pixels.

The raster-scanned display is, inherently, an *X–Y* coordinate display, so the polar coordinates must be converted to rectangular coordinates in order to store the image in the video RAM. This is not a difficult calculation, but it takes time. However, even though the display is updated every $1/50$ s to prevent flicker, the compass card will only rotate a few degrees per second, and orientation of the compass card can be changed a few times per second.

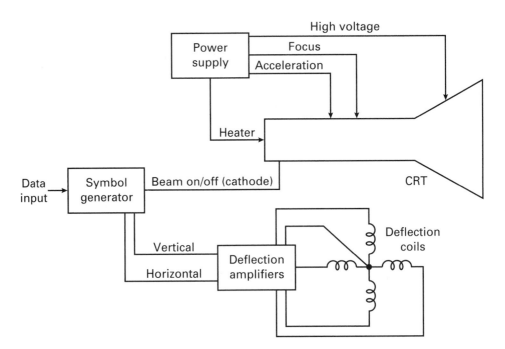

Figure 8–15   Block diagram of an electromagnetically deflected CRT display.

Some attributes of the HSI display do not change regardless of the heading or CDI situation. These include the lubber line, the aircraft icon at the center of the HSI, the glide-slope scale, any alphanumeric legends, and the vertical and horizontal reference lines. These are always present and are retrieved from the ROM and added to the video RAM without any calculations.

In a similar fashion, the course arrow and the CDI are retrieved from the ROM, and the orientation is calculated using the course and VOR information, converted to rectangular coordinates, and added to the video RAM.

The video RAM contains an overlay of the fixed attributes, the compass card, and the course arrow/CDI. The high-speed video RAM provides the information necessary to update the display 50 times per second, while the microprocessor continually recalculates the display picture. Since the fixed attributes do not change, only those attributes that do change are rewritten into the video RAM by the microprocessor.

Although the EFIS can display any form of data, the HSI and the ADI are two of the more common.

The weather radar display processes real-time data from the weather radar unit and stores only a few attributes such as the range rings and alphanumeric information.

ASSOCIATED INSTRUMENTS

Some other instruments need discussion because they are an important part of an electronic system. One of the more important is the *aneroid altimeter.* This instrument measures the altitude above sea level by measuring the ambient air pressure. A special altimeter is used to provide the pressure altitude for the transponder.

The altimeter contains a sealed container, called a bellows, that has corrugated ends; the container is free to expand when the pressure outside decreases such that the air inside causes the bellows to expand. The bellows is connected by a rack and pinion assembly to a pointer that displays the altitude, MSL, as shown in Fig. 8-16. Inserted between the diaphragm and pointer is an adjustable mechanical offset, called the baro set, that provides correction for variable barometric pressure. A knob for setting the sea-level pressure is provided on the front of the altimeter and a calibrated dial that reads the sea-level atmospheric pressure.

An encoding disk is driven from the diaphragm before the baro set and provides an electrical output that is the pressure altitude, which is the altitude that would be read from the altimeter if the baro set were set to 29.92 in. of mercury. This electrical interface is required for the transponder. This type of altimeter is called an *encoding altimeter.* A modification of this altimeter is similar in construction to that shown except that there is no indicator, but only an encoding disk. This type of altimeter is called a *blind encoder.*

The altimeter is mounted in a sealed container and the container is connected by a hose to the outside of the aircraft. This arrangement is called the *static pressure system.* This system is simply a path to the outside of the aircraft that has been carefully arranged to be away from any airstreams so that it provides an accurate example of the ambient atmospheric pressure.

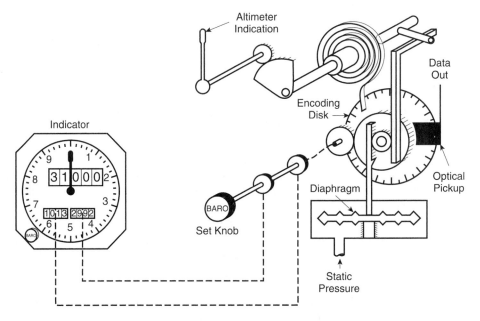

Figure 8–16   Diagram of an aneroid altimeter with an encoding disk.

The orifice to the outside is called the *static port* and is usually placed on the side of the fuselage away from turbulent air, such as that from the propeller, wings, struts, landing gear, or any other items in the slipstream. The static port is fitted with a drain so that water will not accumulate within the port and a heater to prevent the water from freezing before it can drain. It is important that only enough heat be used to melt any accumulated water, because overheating the air will cause altimeter errors.

The rate change of altitude is called *vertical speed,* and an instrument similar to the altimeter is used to measure vertical speed. The *vertical speed indicator,* or VSI, is somewhat like an altimeter with a leak. Imagine an altimeter in which the bellows has a small leak. If the aircraft climbs rapidly, the altimeter will begin to rise, but, because of the leak, the aircraft must climb faster than the leak can equalize the pressure if the altimeter is to continue to show an increase in altitude. Once the aircraft stops climbing, the leak will equalize the pressure and the altitude will return to zero.

The major problem with this simple VSI is that the instrument is very slow to respond and tends to lag behind the actual rate of climb of the aircraft. To achieve a speedup effect, an inertial assist is provided in the form of a suspended weight that helps the pointer move and provide a more accurate indication of vertical speed. This inertially aided indicator is sometimes called an *instantaneous VSI,* or IVSI. It is this instrument that contains the TCAS RA information.

Another important pressure-based instrument is the airspeed indicator. As shown in Fig. 8-17, this instrument measures the pressure differential between the pitot ram air and the pitot static air. Ram air is derived from a probe placed in the slipstream away from tur-

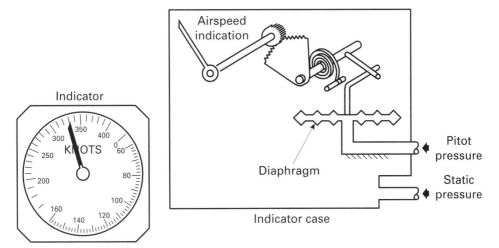

Indicator

Figure 8–17 Airspeed indicator.

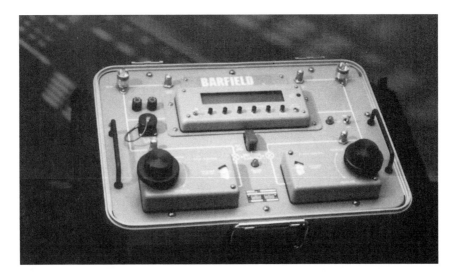

Photo 8–4 The Barfield DPS400 tests aircraft pilot-static systems and is used for leak detection. The unit is also used to calibrate altimeters, airspeed, mach, vertical speed, engine pressure ratio, and manifold pressure indicators. Photo courtesy of Barfield, Inc.

bulence. The differential pressure between the ambient pressure and the pressure produced by the air being forced into the pitot tube provides an indication of airspeed.

The airspeed indicator has a bellows similar to the altimeter, except the bellows is not sealed but is open to the ram air. The airspeed indicator case is sealed and connected to the static port so that the bellows will expand as a function of the pressure difference between the ram air and nominal static pressure. The higher pressure differential due to the

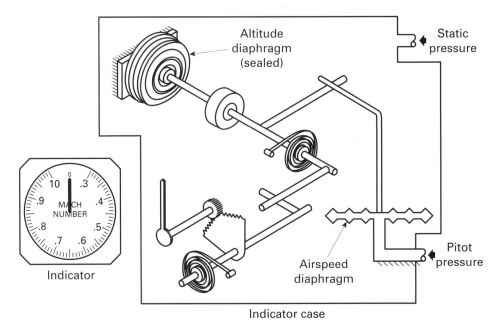

Figure 8–18   Simplified block diagram of a mach indicator.

motion of the aircraft through the air causes the bellows to expand. A similar mechanical connection between the pressure bellows and the indicator provides the indicated airspeed. This is not the true airspeed because corrections are necessary for temperature and altitude.

The final mechanical instrument to be discussed is the *mach indicator.* Airspeed expressed as a mach number is referenced to the speed of sound. However, the speed of sound is a function of the temperature of the air. Instrument makers have found it difficult to use the outside air temperature as the correcting factor, and therefore corrections required to convert airspeed to mach are derived from altitude. Figure 8-18 shows a simplified diagram of a mach indicator. Basically, the mach meter is an airspeed indicator using a diaphragm that senses the pressure differential between the ram air and static air, but a static pressure diaphragm that expands or contracts as a function of altitude provides the necessary correction.

AIR DATA COMPUTER

An air data computer, sometimes called the *central air data computer,* or CADC, is used to provide all parameters relative to the aircraft's position and motion through the air. The computer takes in ram air, static pressure, and total air temperature to provide indicated airspeed, calibrated airspeed, true airspeed, pressure altitude, sea-level altitude, mach, and outside air temperature. The air data computer contains a mixture of mechanical and electrical components to measure and calculate these parameters.

The total air temperature is the temperature of the air surrounding the aircraft, which is above the ambient air temperature because of the compression of the air due to the motion of the aircraft through the air. Compressing air adds energy to the air, and some of that energy is heat. Total air temperature is important because it is the air temperature that enters the jet engines, the temperature of the air striking the leading edge of the wing, which permits or prevents ice formation.

The outside air temperature is the ambient temperature of the air without the effects of heating due to the motion of the aircraft and gives an indication if moisture in the air is ice or water. It is nearly impossible to sample the air in a moving aircraft without compressing it. Therefore, the air is sampled realizing that it is compressed, and corrections are made using the airspeed to derive the actual ambient outside air temperature.

The air data computer senses indicated airspeed, total air temperature, and pressure altitude, three uncorrected parameters. The corrected parameters derived from these, such as mach from airspeed, require corrections based on one of the other two factors. Table 8-2 shows how the parameters are used.

Figure 8-19 shows the block diagram of an air data computer. There are three input parameters that must be converted to electrical signals so that they may be used in the computer to make the necessary corrections.

The pressure altitude is, essentially, an encoding altimeter as explained in Chapter 7. When an air data computer is used, the pressure altitude sensor in the air data computer provides this information to the equipment aboard the aircraft that require it. Typical users of pressure altitude external to the air data computer are transponders, the altimeter, TCAS, and autopilot. Transmitting these data to the required equipment is a perfect example of the use of digital data transmission as discussed in Chapter 4. Using the ARINC 429 system, a single transmitter, the air data computer, provides a number of users with pressure altitude information.

Measuring the total air temperature involves an RTD on a probe mounted on the skin of the aircraft. An RTD is a resistive thermal device, sometimes called a *thermistor*. This device is a variable resistor that changes with temperature. A simple circuit using an A/D converter converts the resistance to a digital number that is used by the microprocessor.

The final input parameter that must be converted to an electrical quantity is indicated airspeed. This is done with a bellows, in the same fashion as with the mechanical airspeed indicator, except an encoding disk is used rather than an indicator dial. The encoding disk is gray encoded, which is transmitted to the computer.

TABLE 8-2

Output parameter	Basic parameter	Correction
True airspeed	Indicated airspeed	Total air temperature Pressure altitude
Outside air temperature	Total air temperature	Indicated airspeed
Mach	Indicated airspeed	Pressure altitude
Vertical speed	Pressure altitude	

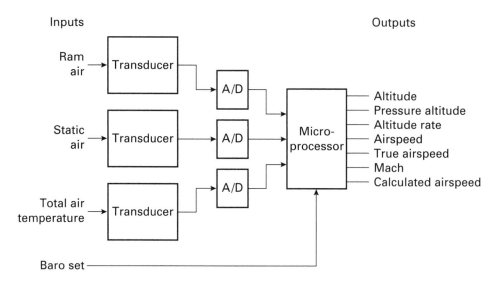

Figure 8–19 Block diagram of an air data computer.

True airspeed is calculated and transmitted to other equipment, such as the autopilot and a true airspeed indicator. Mach is also calculated from indicated airspeed and is used by a mach meter and the autopilot.

REVIEW QUESTIONS

8.1. How does an RMI help the pilot when flying a VOR?

8.2. What information is displayed on an HSI?

8.3. Why is it important to set the OBS to the runway heading in an HSI, in spite of the fact that the OBS has no function on localizer mode?

8.4. How is the CDI indication of an HSI corrected when making an approach to the departure end of a runway using the back course?

8.5. Why is a servo system required for some gyro-driven indicators?

8.6. Where is the glide-slope indicator located when an HSI is used for an ILS approach?

8.7. How does the deflection of the CDI in an HSI differ from the deflection in a OBS–CDI indicator?

8.8. What is a flux gate?

8.9. Why are the signals from the flux gate 800 Hz when the excitation voltage is 400 Hz?

8.10. Why are gyros slaved?

8.11. How are the accelerometers used in an inertial navigation system mounted on an inertial platform?

8.12. What is a vertical gyro used for?

8.13. How does the rate-of-turn indicator gyro differ from other gyros?

8.14. What advantages does the laser ring gyro offer?

8.15. Why were LEDs not readily accepted when they were first introduced into aircraft displays?

8.16. What is sunlight readability?

8.17. What is a pixel?

8.18. If a graphics display is scanned and has 512 lines with an aspect ratio of 1.5:1, how many pixels does the display have?

8.19. If the display of Problem 8.18 is required to provide red, green, blue, and white, as well as dark, how many bits must each pixel contain?

9

Flight Control Systems

CHAPTER OBJECTIVES

In this chapter the student will learn about basic flight control servo systems. The basic division of flight control into three axes will be discussed. The unique problems of microprocessor control of flight critical systems will be covered.

INTRODUCTION

The *flight control system,* often called the *autopilot,* automatically controls the flight of the aircraft to reduce the pilot workload or to increase the accuracy of the aircraft path. As an example, category II and III landings are made with an autopilot because of the inability of a human operator to maintain the necessary flight control precision. A flight control system similar to the autopilot is the *flight director.* This system does not control the aircraft, but gives the pilot the necessary commands to fly a particular situation. Flight directors are used in small aircraft for landing and in larger aircraft as an adjunct to an autopilot.

An autopilot can control as few as one aircraft axis or as many as three. Typically two axes, the roll and pitch, are sufficient for flight maneuvers and the maintenance of altitude. The yaw axis provides assistance on turns.

The autopilot is another example of a servo system. Many components of an autopilot system use servo motors, such as resolvers and the members of the synchro family discussed in Chapter 8. The servo systems in Chapter 8 were used to control the positions of indicators as a function of data from a navigation system.

The purpose of an autopilot is to maintain the flight path of an aircraft according to a predetermined track. To the autopilot, the three control axes are pitch, roll, and yaw, although the end result is a predetermined track in a three-dimensional system: *X, Y,* and altitude. More sophisticated autopilots provide throttle control as well.

For the *X* and *Y* coordinates, the autopilot relies on inputs from a directional gyro or from a radio-based navigation system, such as a VOR or one of the long-distance navigation systems. Some autopilots may be controlled from either the ILS or MLS landing systems for landing control. In many cases, the RNAV is the major supplier of autopilot information. The vertical axis may be controlled from an altimeter or from glide-slope information from a landing system. The autopilot provides motion to the control surfaces of the aircraft—the ailerons, elevator, and rudder—to provide control in the roll, pitch, and yaw axes.

The autopilot consists of several subassemblies. First, a control panel is mounted on the flight deck either in the instrument panel or in a pedestal that contains switches for engaging and disengaging the autopilot and setting the mode of operation. The second element is the autopilot computer, which accepts the flight information from other flight systems and the settings of the control panel, processes these data, and provides outputs for the servo motors that control the aircraft control surfaces.

The third element of the autopilot is the servo systems that drive the control surfaces—the ailerons, elevator, and rudder. The control surface servo systems are either electrical or hydraulic, depending on the nature of the control system in the aircraft. Smaller aircraft have electrical servos; here electric motors are coupled to the controls of the aircraft. In larger aircraft, the autopilot provides hydraulic valve signals that move the control surfaces.

The autopilot is divided into three *channels,* roll, pitch, and yaw. These channels operate relatively autonomously with only a small amount of signal sharing. In some modes of operation, only one or two of the three channels operates. For clarity of understanding, the three channels will be investigated individually.

ROLL CHANNEL OPERATION

Roll stabilization requires input from the vertical gyro roll channel synchro, as shown in Fig. 9-1. When the autopilot is not engaged, a synchro in the autopilot roll channel, S_1, is driven by a servo motor in a closed-loop servo system such that it follows the roll output from the vertical gyro. When the autopilot is engaged in the roll attitude hold mode, the servo motor is de-energized such that the synchro remains in its last set position. The rotor of the synchro is used as a source of error signal for the autopilot computer. This error voltage drives the servo for the ailerons, which will adjust the roll of the aircraft such that the output of the now stationary roll channel synchro is at null. Thus, the aircraft will continue the angle of roll at the time of engagement.

Many autopilot roll computers contain a resolver that is mechanically coupled to the roll channel synchro. This resolver is arranged such that it provides a null output when the synchro is at 0° or vertical. When the wings level mode is selected, the motor drives the

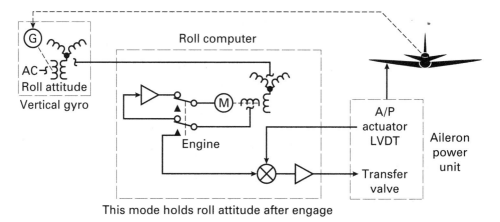

This mode holds roll attitude after engage

Figure 9–1   Block diagram of the roll channel.

resolver and the synchro together until both are at 0°, which is characterized by the null from the resolver. The output of the roll channel synchro is used as an error signal for the aileron servo system. This causes the aircraft to change its roll angle until the error signal from the roll channel synchro is zero. This zero error signal occurs when the vertical gyro and the roll channel synchro are at the same position, which is 0°, or vertical.

The autopilot includes a *turn knob*. This control, typically a knob, is used to set a specific turn rate from the autopilot control panel. More accurately, it sets a roll angle, which in turn sets a rate of turn. When the turn knob is removed from its detented, neutral position, the ac voltage from the turn knob causes the servo motor to change to the position of the computer roll angle synchro. Since this can be done while the autopilot is engaged, a signal from the rate gyro is introduced to limit the rate of turn.

PITCH CHANNEL OPERATION

Pitch information is obtained from the vertical gyro pitch channel synchro in a fashion similar to the roll channel synchro. The pitch channel in the autopilot follows the position of the gyro pitch synchro using a servo loop as explained for the roll channel.

For the pitch hold mode of the autopilot, at engagement the servo motor used to drive the pitch synchro in the autopilot is disconnected, thus causing the pitch synchro to remain stationary. The error signal from the autopilot pitch synchro is used to drive the elevator servo, which adjusts the pitch of the aircraft to equal that set before engagement.

One of the more common autopilot modes involving the pitch channel is altitude hold. In this mode, an error signal from an altimeter or air data computer is introduced to control the pitch synchro. If the altitude is to be increased, the error signal from the pitch synchro drives the elevators until there is greater nose-up attitude and thus a null from the synchro rotor. As the altitude approaches the desired level, the amount of pitch forced into the pitch synchro is reduced and eventually returns to level flight upon reaching the desired altitude.

Other inputs are used to control the pitch of the aircraft. The autopilot may be coupled to ILS or MLS receivers during an approach, and the descent may be controlled. The operation of these modes is similar to the altitude hold mode, but error signals from the glide-slope receiver or MLS receiver are substituted for the altitude error.

YAW CHANNEL OPERATION

The pitch and roll motions of an aircraft cause the aircraft to turn and change altitude. Although the aircraft yaw angle changes during maneuvers, these motions are usually a result of the turn.

The yaw control alleviates undesired yaw due to various forces on the aircraft. Basically, an aircraft is designed to fly straight and level. A wind gust may cause the aircraft to yaw, but once the gust has ended, the natural stability of the aircraft returns the aircraft to straight and level. However, the aircraft will tend to oscillate as it is buffeted by gusts. The yaw channel is used as a damper to remove these oscillations and thus is often called a *yaw damper.* Another term for this mode of operation is *stabilization augmentation system,* or SAS.

The output from the rate-of-turn gyro is used as a sensor input for the yaw damper. The ac output of the resolver is demodulated, and the resulting dc is high pass filtered, as shown in Fig. 9-2. The result is a signal that represents the acceleration of the yaw angle. The frequency range of the accelerations from the high-pass filter is within the range of the natural resonant frequency of the aircraft.

The acceleration signal is used to drive the rudder servo system, which causes the rudder motion to counter the yaw due to wind gusts. To improve the operation of the yaw damper, airspeed data from the air data computer are used to adjust the gain of the yaw channel.

One characteristic of the aircraft, the resonant frequency of the aircraft yaw, is an important part of the design of the yaw channel. Airspeed is introduced to further improve the dynamic characteristics of the autopilot. It is important to understand that all three chan-

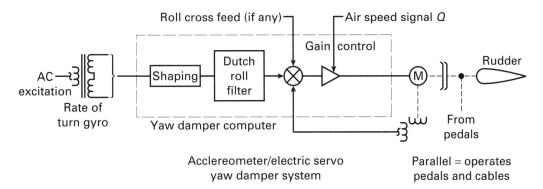

Figure 9–2 Block diagram of the yaw channel.

nels of the autopilot are tailored to the dynamics of the aircraft. An autopilot is designed to be used in a specific aircraft. Autopilots used in smaller aircraft may be programmed to match the dynamics of a specific aircraft type.

AUTOPILOT SAFETY

The autopilot is a very crucial aircraft electronic system because of its interface with the actual control surfaces of the aircraft. If part of an autopilot fails, such as a servo system, there is a possibility of erratic operation of the flight control surfaces. The most dangerous form of failure is *jamming* of a control surface due to an autopilot component. This form of failure is prohibited by providing slip clutches, breakaway pins, and other safety mechanisms that allow the jammed surface to be freed. Very careful design is another technique used to avoid this type of catastrophic failure.

Another failure mode is the *hardover.* This failure causes the servo motor of an autopilot to drive the servo to the limits of its travel. This type of failure can be sensed electrically and the autopilot can be disengaged. If the failure evades detection, hardover will occur. To prevent hardovers from causing violent maneuvers, the rates of the servo systems are limited, so the chance of the hardover being stopped manually by the pilot or copilot before the aircraft has assumed a dangerous attitude is good.

The worst type of hardover is the multiple-axis hardover. This means that the aircraft has violent maneuvers in two or three axes. The danger is that the aircraft will be placed in a critical attitude before the autopilot can be disengaged. To prevent this type of failure, the three autopilot channels must be as independent as possible. This independence is achieved by providing separate computing circuits and input/output channels, three independent power supplies, and so on.

The use of microprocessors in autopilots was avoided for many years because of the failure modes of the microprocessor. The general mistrust of the microprocessor was due to the possibility of the processor becoming glitched and running amok. Because of this, the failure mode of the processor was completely unpredictable. The autopilot was not the only application of the microprocessor that could suffer from unpredictable failures. Microprocessor applications in automotive electronics, industrial process controls, and others drove research and development into fault-tolerant software and hardware monitoring circuits. Eventually, software and hardware were improved so that the microprocessor is suitable for use in autopilots.

The microprocessor has not been developed to the point where it will never fail; the likelihood of a failure is just very remote. However, the dangers associated with a multiple-axis hardover are so great that autopilot designs often contain multiple microprocessors so that the failure of a microprocessor will produce only a single-axis hardover. Each processor has a monitor circuit, sometimes called a *watchdog,* that continually checks the operation of the microprocessor.

Watchdog circuits can be implemented with a number of methods, but usually the microprocessor is required to set an external circuit at a regular rate. If the circuit is not set, it "times out" and disengages the autopilot. The setting of the watchdog can be the result of a simple statement in the program or the result of a test calculation. The test calculation is more effective than the simple statement, because if the computer is in a bogus loop that happens to include the "watchdog set" statement, the watchdog timer will be reset and the autopilot will not be disengaged. If the watchdog is set as a result of a simple routine that involves writing to memory or jumping to a subroutine, it is much less likely that a microprocessor caught in a short loop will be able to execute this test routine and successfully set the watchdog timer.

The introduction of electronic circuits to replace the motorized synchros and other electromechanical components used in autopilots had a similar rocky road to acceptance. Newer autopilot designs use digital storage techniques to hold heading and altitude information. The electromechanical servo systems were unreliable, heavy, power hungry, and expensive. The introduction of these functions into the microprocessor provides a reliable and very inexpensive replacement for electromechanical components.

The roll, pitch, or yaw channel synchros are memory devices. As an example, the roll channel synchro follows the vertical gyro when the motor is engaged. When the roll attitude hold is engaged, the motor is disabled, and the synchro is stationary and is essentially a memory device. An alternative method of providing this function is to convert the output of the vertical gyro synchro to a digital number with an analog to digital (A/D) converter. The digital number can be stored in computer memory. When the autopilot is energized in the roll attitude hold mode, the output of the vertical gyro roll synchro continues to be converted to a digital number, and the number is compared to the number in the memory. The error signal is the difference between the current converted value of the vertical gyro roll synchro and the number stored in memory. This error signal is used to generate the drive for the aileron servo system by using a digital to analog (D/A) converter to provide a 400-hertz (Hz) drive signal.

The replacement of synchros and resolvers has been a desire of avionics systems engineers for some time. These devices are very expensive and are quite heavy. However, because of their importance in airborne systems, these servo motors have been refined to a very high state of precision, and with the use of automated manufacturing the costs have been kept under control.

The purpose of a synchro or resolver is to convert mechanical shaft position to an electrical signal, an analog signal. When a digital autopilot is employed, this analog signal is converted to a digital data word. If a digital shaft angle sensor could be used, the A/D converter could be eliminated. A digital shaft angle sensor is called a *shaft encoder*. A well-designed synchro or resolver coupled to a good A/D converter can provide angular precision to small fractions of a degree. A shaft encoder that can supply a digital angle resolution equal to that of a resolver or synchro represents a product of a mature technology, and it took some time before resolvers and synchros were replaced with shaft encoders.

Digital shaft encoders can be made with very low inertia and friction. Later-technology shaft encoders have reliability and performance that rival the best servo motors, and as shaft encoder production technology matures, the cost of shaft encoders will be competitive with servo motors.

A digital autopilot consists of digital sensors, shaft encoders connected to the vertical and directional gyro, digital data from navigation equipment sent by ARINC 429 or

ARINC 629, and digital signal processing. The output of the autopilot that positions the aircraft's control surfaces is usually analog, because the aircraft is an analog device! A technique of improving the reliability of an autopilot, which is applicable to both analog and digital systems, but is applied almost exclusively to digital, is fault tolerance.

In previous discussions of autopilot safety, techniques of minimizing and detecting failures were discussed. Our discussion implied that it is important that the electronic systems used for flight critical systems should not fail, and when they do fail, the failure should be detected and the system disabled. In the case of an autopilot, a detected failure alerts the crew, disengages the autopilot, and does not cause a violent maneuver. Although this situation represents a failure of a critical system, the aircrew can safely control the aircraft. This situation is called *fail safe*.

In many cases, redundant elements are provided so that errant systems may be replaced with a backup system and the autopilot restored to operation. If an autopilot contains redundant systems and the failure of an element is detected and the failed system is automatically replaced with a backup such that the autopilot system operates continuously without fail, the technique used is called *fault tolerance*. In this situation, parts of a system can fail while the entire system continues with no degradation of performance.

There are several methods of achieving fault tolerance. One method is *redundancy*. This involves two or more systems operating continuously. One system provides the autopilot action, while the second system is ready to be placed into operation should the first system fail. The important part of this technique is to determine if a system has failed. Simple methods of failure detection, such as the watchdog, do not provide an accurate indication of a system failure for a flight critical system such as an autopilot.

One effective method of determining a system failure in a redundant system is to have both systems fully operational and compare the outputs of the two autopilot systems. If the outputs differ by a predetermined amount, one system has failed. The difficult part of a dual redundant system is to determine which system has failed. A method of doing this is to perform diagnostic test routines for the two redundant autopilot systems to determine which system has failed.

The diagnostic routines must be performed quickly because control of the aircraft must be restored as quickly as possible. Diagnostic tests are not guaranteed to find the errant system. The more effective a diagnostic routine is, the longer it takes for the computer to run the diagnostic. Although long, effective diagnostics can find most failures, the autopilot cannot be out of order for as long as it may take to find the failed system.

A superior method of achieving fault tolerance is to operate three redundant systems in parallel and to compare the three outputs using a technique called *voting*. This technique shuts one system down if two systems have identical outputs with the third different. It is possible that two systems can fail simultaneously, leaving the third system as the sole operating system. This would leave the two errant systems operating and the not-failed system shut down, thus causing a failure. The likelihood of two systems failing simultaneously is very remote if the systems are designed correctly.

It is very important that the voting circuits have extreme reliability because it is up to the voting circuits to determine which of the three systems has produced a different result. If a voting circuit does not detect a different result, the failed system will go undetected and may provide a false output. On the other hand, if a voting system detects a different output when there is none, one of the three systems will be disabled unnecessarily. This loss of one of the three systems does not cause the system to fail but renders the system no longer fault tolerant. A real failure cannot be detected because there are no longer three systems to compare.

It was suggested that two systems cannot fail simultaneously. This is not totally true. If the systems are identical, it is possible that a set of input conditions can cause two or even all three systems to fail. This is particularly possible if there is a software failure. To prevent this type of failure, the three operating systems use different software and hardware designs.

AUTOPILOT SERVO SYSTEMS

The servo systems used for controlling the ailerons, elevator, and rudder of the aircraft are a function of the type of flight control systems used on the aircraft. In smaller aircraft in which the aircraft control surfaces are directly connected to the control yoke through cables, the autopilot servo motors are physically connected to these mechanical systems. The servo motors have an electric clutch that will physically connect or disconnect the servo motor from the aircraft control system. Therefore, when the autopilot is not used, the motors will not add drag to the aircraft control system. When the autopilot is not operating, it is said to be disengaged.

An autopilot servo system contains three elements, the amplifier, the motor, and a position sensor. The position sensor is part of a feedback circuit that causes the servo motor to move the control surfaces at a precise speed and prevents the servo system from causing the control surfaces to be fully deflected. When the surfaces are fully deflected, the servo motor will attempt to push the surface beyond its normal mechanical travel, which can cause damage to the servo or the flight control surfaces.

As the autopilot is controlling the aircraft, the control yoke moves in response to the autopilot commands. To prevent damage to the system should the control yoke be restrained while the autopilot is engaged, the servomotor includes a slip clutch to limit the maximum amount of torque available from the autopilot.

The motors used for the control surfaces may be ac, dc, permanent magnet, or electromagnet, and they may or may not include gear reduction drives. The motor is not required to provide a significant average power, but may be needed to provide bursts of high torque, particularly from stall. The motor with the best-suited torque characteristics for a light plane autopilot is the permanent magnet dc motor.

This motor requires a bidirectional dc driver because the motor must turn in both directions. Typically, a servo amplifier is used to drive the motor where the amplifier is of the "bridge" type, as shown in Fig. 9-3. The advantage of the bridge amplifier is that bidirectional current may be applied to the motor without generating a negative supply voltage. The disadvantage is that two wires must be supplied to the motor. This prevents one side of the motor from being connected to ground and requires that two large-gauge wires be used, thus increasing weight and power loss through the wires.

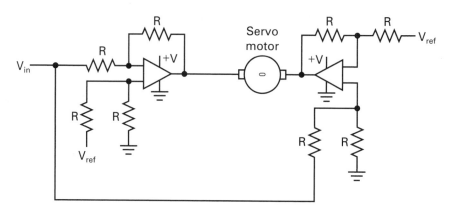

Figure 9–3 A bridge amplifier for supplying bidirectional current from a single polarity power supply.

Servo motors can require a significant amount of current to provide the bursts of torque required for autopilot operation. To provide this current, large wire must be provided for the motors. The problem with this arrangement is that heavy pulsed current may be required to run a considerable distance. An improvement may be made by mounting the servo amplifiers at the servo motor, and there the heavy pulsed current only needs to travel a short distance from the amplifier to the motor. The disadvantage of mounting the amplifier on the motor is that the amplifier is in an unheated, unpressurized, and often inaccessible location.

Autopilots can be coupled to any of the common radio navigation systems. For en route use, the VOR, LORAN-C, Omega, and GPS sensors may be used to control aircraft flight. With the exception of GPS, altitude hold is used to control the aircraft altitude. For landing, the ILS or MLS may be coupled to the autopilot. Before any radio navigation system can be coupled to the autopilot, the navigation signal must meet various criteria.

The first criteria is that the radio navigation system to which the autopilot is to be coupled may not be flagged, but must have a workable signal. Second, the navigation sensor must not be at full scale, but somewhere in the operating range of the system. As an example, if an aircraft is far left of a runway center line such that the localizer indicator is at full scale right, an autopilot will not allow coupling to that navigation system until the aircraft is closer to the center line, and the normal meter indication is off the left stop. The autopilot is capable of holding altitude, heading, or courses, but not suited for finding positions from a distance. When the autopilot has determined that the radio navigation is close enough to intercept and follow, the autopilot is said to have *captured* the radio navigation aid and will successfully follow the guidance information.

CONTROL PANEL

The autopilot control panel contains controls for setting the mode of the autopilot and for controlling the rate of turn and the rate of climb or descent. Figure 9-4 shows a typical control panel. This panel contains two engage switches that transfer the control of the aircraft from manual pilot control to automatic control. These switches are magnetically held; that

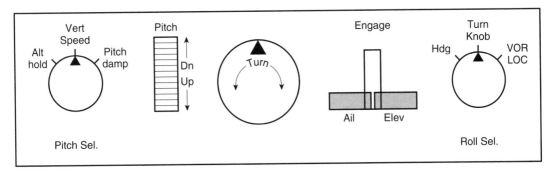

Figure 9–4   Diagram of an autopilot control panel.

is, if the autopilot circuits and the navigation sensors are fully functioning, the autopilot will allow engagement. Should a navigation sensor lose signal or should there be a failure in the autopilot, the holding electromagnets are de-energized and the switches fall to the disengaged position. In this example, the pitch and roll channels have separate engage switches and are labeled elevator and ailerons, elev and ail, respectively.

In the center of the control panel is a turn knob. This knob is used for manually initiated turns and is active only when the roll mode selector is in the turn-knob position. When the turn knob is in the neutral position, the aircraft flies the heading that existed when the autopilot was engaged.

The equivalent control for the pitch channel is to the left of the turn knob and is the vertical speed selector. Like the turn knob, this selector is active only when the pitch channel selector is in the vertical speed position. If the vertical speed knob is in the neutral position, the aircraft will be in an altitude hold mode when the autopilot is engaged.

The pitch channel mode selector is the left side control and selects between the vertical speed knob and pitch damp. Pitch damp is a function similar to yaw damp and is used when the aircraft is flown in turbulence.

On the right side of the control panel is the roll channel mode selector. When this mode selector is in the heading select position, the heading as set, usually on the HSI, controls the aircraft. When the mode switch is in the turn-knob position, the aircraft holds the heading while the autopilot is engaged and may be changed by the turn knob. When the switch is in the VOR/LOC position, the aircraft will fly a VOR or localizer beam depending on which navigation facility is selected by the navigation radio receiver. The final mode is ILS, which is allowed only when a localizer frequency is set on the companion navigation radio. This mode causes the roll channel to follow the localizer beam, and also causes the pitch channel to follow the glide-slope beam. Additional selection of navigation sources (LORAN-C, GPS, and so on) is accomplished external to the autopilot control panel.

AIR DATA COMPUTER

The air data computer provides information for various flight instruments and for the autopilot. This information is derived from the dynamic and static parameters of the outside air. The air data computer determines the following parameters:

1. *Total air temperature,* TAT. This parameter is the temperature of the outside air, which includes the normal rise of temperature due to the ram effect. When air is compressed, as it is when it enters the pitot tube, the temperature of the air is increased. Since it is difficult to sense air temperature without some compression, TAT is measured and corrections are made as a function of airspeed.

2. *Pressure altitude.* This parameter is simply read from an aneroid altimeter without correction for atmospheric pressure. Pressure altitude is used for the radar transponder and may be used for altitude hold, particularly for high-altitude flight.

3. *Indicated airspeed.* This parameter requires ram air from the pitot tube and provides information for an autopilot and for panel-mounted indicators.

4. *Calibrated airspeed.* This parameter is used for indicators, as well as for the autopilot. The total air temperature and altitude are used as correction factors for determining calibrated airspeed.

5. *Vertical speed or rate of climb.* Differentiating the altitude as a function of time in the computer provides the vertical speed parameter. In older mechanical instruments, this differentiating was accomplished by providing a chamber with a calibrated leak. If the atmospheric pressure at altitude changed faster than the calibrated leak could equalize it, the pressure built up in a diaphragm, which caused a pointer to deflect and indicate vertical speed. Such a device was called a *vertical speed indicator,* or VSI. Because the calibrated leak was a slow process, the indicated vertical speed tended to lag the actual vertical speed. To improve the response time of the vertical speed indicator, a suspended mass was used to sense acceleration and aid in moving the vertical speed indicator. This form of VSI has a much faster reaction time and is often called an instantaneous vertical speed indicator, or IVSI. The air data computer provides the instantaneous vertical speed output for use by the VSI or IVSI and the autopilot.

6. *Mach.* This parameter is a measure of speed based on the velocity of sound at the altitude and temperature at which the aircraft is operating. The calculation of mach involves the TAT and the indicated airspeed. The mach output is provided for the mach indicator and for the autopilot for use in the mach hold mode.

AVIONICS SYSTEMS

Throughout this text, we have discussed navigation systems, display technologies, autopilots, and communications systems. Historically, navigation has been accomplished using one type of navigation system. Typically, an aircraft would be flown by VOR, LORAN-C, or maybe an ADF beacon. Often a pilot employed a second navigation system to enhance the navigation accuracy. This involved, perhaps, two different VOR stations or a VOR and an ADF together.

Each type of navigation system has its strengths and weaknesses. The ADF has problems at night, the VOR suffers from multipath propagation in mountainous areas, and the LORAN-C accuracy is not as good as VOR when the VOR does not suffer from multipath propagation.

Why not use every type of navigation equipment in the aircraft and use the navigation signal that offers the best precision at the time? As the flight progresses, the navigation might involve VOR, DME, GPS, LORAN-C, Omega, and so on. When several navigation signals are strong and accurate, their results can be averaged to improve the navigation accuracy or used to evaluate the signal integrity of other navigation systems.

In addition to the radio navigation equipment, inertial navigation equipment, gyros, and the air data computer may all provide additional information to aid the navigation.

Remember the HSI; this instrument attempts to present a picture of the navigation situation. With electronic indicators, navigation data can be combined with weather information, collision avoidance, and waypoint locations to create the ultimate HSI.

Furthermore, why not couple this navigation system to the flight control system and, effectively, have an automatically controlled aircraft? There are no reasons not to create this ultimate navigation and control system and reap the benefits of the precision navigation accuracy available from complete electronic control.

To achieve this goal, several avionics systems are required. First, an electronic display is necessary to display the complex data that will be available. No mechanical instrument is capable of displaying the amount of data available. Second, an effective data communications system is required to transmit data from each navigation system (VOR, DME, LORAN-C, and others) to be processed. Finally, a computer is required to control the system, input the desired waypoints, and solve the navigation problem.

This navigation system is rapidly becoming the architecture of modern electronic navigation. What were stand-alone navigation systems, such as the VOR and DME, are now called *sensors*. These sensors may have only limited data-processing capability and provide only raw data to the navigation computer to be processed and used to solve the navigation problem.

The heart of a modern navigation system is the navigation computer. This computer is provided with user-entered data as waypoints, altitudes, flight plans, offsets and so on. The computer also receives data by data communications from the sensors. The computer's main task is to solve the navigation problem. Essentially, the navigation problem is how we get there from here. This is no new situation. The VOR system provided steering information such that, if the aircraft was turned to correct for course errors, the aircraft would fly to the VOR. However, much more information may be desired, such as how much distance remains to the waypoint, what is the ground speed, and when will the aircraft arrive at the waypoint.

Data communication is of extreme importance in a modern aircraft navigation system. Clearly, data are required from each sensor. The computer will provide frequency tuning for those radio navigation systems that operate on multiple frequencies, and therefore, for some radio navigation systems, two-way data are required. These important data highways can use one of the two common ARINC characteristics, ARINC 429 and ARINC 629. A significant amount of data needs to be transmitted in the modern navigation system, and the higher speed of ARINC 629 is an advantage. Fiber-optics transmission of the ARINC 629 is also an advantage in large avionics systems.

The display includes the waypoints, the desired tracks between waypoints, weather information, and proximate aircraft from the TCAS equipment. This type of display is called a *moving map display* and is the ultimate in horizontal situation indicators. By coupling the autopilot to the navigation computer, the aircraft will automatically fly the desired flight, while automatically switching waypoints.

INTEGRATED MODULAR AVIONICS

We have developed the modern aircraft electronics system from the individual, autonomous navigation system to the navigation computer and the fully integrated navigation system with its sophisticated communications and displays. The final step is to introduce the most modern concept of navigation systems, the integrated modular avionics system.

The concept of the LRU, the line replaceable unit, was a subassembly that could be replaced in the field, but still contained a considerable amount of circuitry. The VOR receiver, as an example, is a sensor that is usually an LRU and is always found as a part of a navigation system.

Using the normal concept of LRUs produces a number of duplicated items. As an example, every LRU has a power supply, a housing or case, connectors, cooling, and so on. Also, since the LRU is separate from the other elements of the navigation system, significant communications is required. Many LRUs have computers for controlling the unit or processing some of the received information.

Some common items may be shared, such as the power supply. A larger power supply that provides the needed internal operating voltages could be built for all the LRUs of a system. The danger in making one power supply is that a failure of that power supply can knock out the entire avionics system. On the other hand, the one power supply could be more reliable than individual power supplies by employing a high-quality design.

One of the more failure prone items of a large electronic system are the cables and connectors. By mounting all the subassemblies of an avionics system in a common enclosure, called a *cabinet,* much of the vulnerable cables and connectors are eliminated. In addition, the cost of individual housings has been eliminated.

The individual subsystems are configured as printed circuit boards, which requires plug-in boards. Although there are still plugs and interconnecting devices, since they are mounted within a housing, much of the dirt, contamination, and handling damage that occurs to external connectors is reduced, and the resulting reliability is vastly increased.

Communications that took place with serial data transmission and multiplexing to reduce the amount of wires and to it protect from electromagnetic interference can now be achieved with more reliability because of the much shorter distances involved and the shielding effects of the avionics cabinet. Because of the short distances involved, parallel bit transfer can be used and thus data throughput is vastly increased.

The semiconductor industry has produced incredibly high performance microcomputers with very high clock frequencies. These chips can process so much data that one computer can process data for several subsystems. The high-speed clocks that drive fast computers are also a source of problems, particularly with radio-based subsystems. Because of this, the plug-in modules of the integrated modular avionics systems must be shielded from radiated interference.

Because of the possibility of single-point failures, it is imperative that integrated modular avionics employ fault-tolerance techniques. But, in spite of the problems, the integrated modulator avionics concept has merit and is becoming a method of avionics integration.

REVIEW QUESTIONS

9.1. What are the three autopilot channels?

9.2. What is a hardover?

9.3. How are microprocessors protected from glitch-induced failures?

9.4. How does the turn knob function?

9.5. What are some of the reasons that an autopilot would be automatically disengaged?

Glossary

The aviation industry has been notorious for its use of abbreviations and acronyms. Although most abbreviations and acronyms that are used in this text are defined at their introduction, they are often used again throughout the book.

These have been included in the glossary for convenience, in addition to some important terms.

AC	*Advisory Circular.* A publication of the FAA which provides important information on a variety of subjects.
AD	*Airworthiness Directive.* A notice of possible safety-related defects in an aircraft.
ADF	*Automatic Direction Finder*
ADI	*Attitude Direction Indicator*
AGL	*Above Ground Level*
AILS	*Automatic Instrument Landing System*
AIM	*Airman's Information Manual*
ALT	*Altitude, Altimeter*
AOG	*Aircraft on Ground.* Used to mark orders and shipping containers for parts required for aircraft that are grounded until the arrival of the part.
ARTCC	*Air Route Traffic Control Center.* The facility that has control over en route traffic in controlled airspace.
ASR	*Airport Surveillance Radar.* A short-range primary radar.
ATC	*Air Traffic Control.* Often refers to ATCRBS

ATCRBS	*Air Traffic Control Radar Beacon System*
ATIS	*Automatic Terminal Information System*
BC	*Broadcast.* Referring to a broadcast station received on an ADF receiver, also refers to the transmission of a mode-S message meant for general information.
	Back Course. The localizer signal in the region of the takeoff end of an instrument runway.
CAT I	Category I, and II, and III designators for the precision of ILS landings.
CAT II	
CAT III	
CDI	*Course Deviation Indicator.* Used with localizer, glide slope, and VOR.
Comm or Com	Having to do with communications, usually VHF communications, such as Comm receiver, Comm antenna, etc.
DF	*Direction Finder.* Usually a manually operated unit, as distinguished from an automatic direction finder, or ADF. Also refers to a ground-based VHF direction-finding system for emergency use.
DG	*Directional Gyro*
DH	*Decision Height.* Associated with instrument landings.
DME	*Distance Measuring Equipment*
DOT	*Department of Transportation.* A U.S. Government agency.
DVOR	*Doppler VOR*
EADI	*Electronic Attitude Direction Indicator*
EFIS	*Electronic Flight Instrument System*
ELT	*Emergency Locator Transmitter*
FAA	*Federal Aviation Administration.* A U.S. Government agency.
FAR	*Federal Aviation Regulations*
FBO	*Fixed Base Operator.* Services offered at an airport.
FCC	*Federal Communications Commission.* A U.S. Government agency.
FL	*Flight Level.* A system of altitude definition using an altimeter not corrected for sea-level barometric pressure.
FMS	*Flight Management System*
FSS	*Flight Service Station*
GADO	*General Aviation District Office.* A local office of the FAA concerned with the regulation of general aviation.
GDOP	*Geometrical Dilution of Precision*
GMT	*Greenwich Mean Time* (or "zulu" time)
GND	*Ground.* The potential of the airframe in an aircraft electrical system; also ground control at an airport.
GPS	*Global Positioning System*

GS　　　　　　*Glide Slope.* Also *Ground Speed.*

HF　　　　　　*High Frequency.* The frequency range from 3 MHz to 30 MHz, usually referring to a single sideband communications transceiver operating in the 3- to 30-MHz range.

HSI　　　　　*Horizontal Situation Indicator*

HUD　　　　　*Heads Up Display.* A system that projects electronically generated instruments on the aircraft wind screen.

IAS　　　　　*Indicated Air Speed*

ICAO　　　　*International Civil Aviation Organization*

ID　　　　　　*Identification* (see IDENT)

IDENT　　　　*Identification.* The Additional pulse added to a radar transponder reply (see Chapter 6). The Morse code identification applied to various navigation services, such as ILS, VOR, and DME.

INS　　　　　*Inertial Navigation System*

IVSI　　　　　*Instantaneous Vertical Speed Indicator.* A vertical speed indicator with an accelerometer to decrease the response time of the instrument.

kHz　　　　　*Kilohertz.* 1000 cycles per second.

kts　　　　　*Knots.* One nautical mile per hour.

LAT　　　　　*Latitude*

LAT/LONG　　*Latitude/Longitude.* Refers to navigation information referenced to latitude and longitude.

LF　　　　　　*Low frequency.* The frequency range from 30 kHz to 300 kHz. Usually refers to nondirectional beacons, and often, incorrectly, to beacons above 300 kHz.

LMM　　　　　*Locator, Middle Marker.* Refers to the 75-MHz beacon on an instrument approach.

LOM　　　　　*Locator, Outer Marker.* Refers to a 75-MHz beacon on an instrument approach.

LORAN　　　　*Long-Range Aid to Navigation.* Refers to several systems of radio hyperbolic navigation systems, usually LORAN-C.

MF　　　　　　*Medium Frequency.* The frequency range from 300 kHz to 3 MHz. This frequency range contains nondirectional beacons and standard broadcast stations that are used for navigation by the ADF.

MLS　　　　　*Microwave Landing System*

MSL　　　　　*Mean Sea Level.* The true altitude above sea level.

NAVAID　　　*Navigation Aid.* Refers to any one of the radio navigation aids.

NDB　　　　　*Nondirectional Beacon.* Refers to a low-frequency or medium-frequency transmitter that is used for navigation with an ADF.

NM (NMI)　　*Nautical Mile.* 6076.1 feet.

NOTAM   *Notice To Airmen.* Notices concerning flight safety transmitted by various means to pilots.

NTSB   *National Transportation Safety Board.* An agency of the U.S. Government that investigates aircraft mishaps.

OBS   *Omni Bearing Selector*

OM   *Outer Marker.* A 75-MHz beacon on an ILS approach.

PAR   *Precision Approach Radar*

PWI   *Proximity Warning Indicator.* A collision-avoidance or terrain-avoidance system indicator.

RA   *Resolution Advisory*

Rad Alt   *Radar Altimeter*

RADAR   *Radio Detection and Ranging*

RBN   *Radio Beacon.* Usually refers to a nondirectional beacon.

RCV   *Receive*

RCVR   *Receiver.* Usually refers to a communications receiver.

RNAV   *Area Navigation Computer*

RVR   *Runway Visual Range.* The visibility at the runway location.

TA   *Traffic Advisory.*

TACAN   *Tactical Air Navigation*

TAS   *True Air Speed.* The indicated air speed corrected for errors.

TCA   *Terminal Control Area, now called class B airspace.* The airspace surrounding major air hubs in the United States and that requires, for entry, additional pilot and aircraft qualifications.

TCAS   *Traffic Alert and Collision-Avoidance System.* A collision-avoidance system.

TRACON   *Terminal Radar Approach Control.* An airport radar-based approach system.

TRSA   *Terminal Radar Service Area*

TSO   *Technical Standard Order.* Government-issued standards for avionics equipment and other aircraft components.

TVOR   *Terminal VOR.* A low-power VOR or VORTAC for short-range use at an airport.

TWEB   *Transcribed Weather Broadcast.* Weather information available on a nondirectional beacon or VOR, or by telephone.

UHF   *Ultra High Frequency.* The frequency range from 300 MHz to 3000 MHz, which includes glide-slope and DME radio navigation systems.

UNICOM   *Universal Communication Frequency*

VASI   *Visual Approach Slope Indicator.* A system of lights that provides the pilot with a visual glide slope.

VFR	*Visual Flight Rules*
VHF	*Very High Frequency.* The frequency range from 30 MHz to 300 MHz, which includes VOR, marker beacons, and communications.
VOR	*Very-High-Frequency Omnirange.* The most common form of air navigation.
VORTAC	A combination of a VOR and a TACAN ground station.
VOT	*VOR Test Facility.* A calibrated VOR signal used for testing a VOR system.
VSI	*Vertical speed indicator*
WAC	*World Aeronautical Chart.* A 1:500,000-scale aeronautical chart.
WP	*Waypoint.* A location en route in an aircraft trip.
WXR	*Weather Radar*
Z	Greenwich Mean Time, or "zulu" time.
ZM	*"Z" Marker.* A 75-MHz marker for en route navigation; now practically obsolete.

Index